T0310402

Forest structure, function and dynamics in Western Amazonia

Forest structure, function and dynamics in Western Amazonia

Edited by
Randall W. Myster

WILEY Blackwell

Registered Office: John Wiley & Sons Ltd, The Atrium, Southern Gate, Chichester, West Sussex, PO19 8SQ, UK

Editorial offices: 9600 Garsington Road, Oxford, OX4 2DQ, UK
The Atrium, Southern Gate, Chichester, West Sussex, PO19 8SQ, UK
111 River Street, Hoboken, NJ 07030-5774, USA

For details of our global editorial offices, for customer services and for information about how to apply for permission to reuse the copyright material in this book please see our website at www.wiley.com/wiley-blackwell.

Library of Congress Cataloging-in-Publication Data

Title: Forest structure, function and dynamics in Western Amazonia / edited by
Randall W. Myster.
Description: Chichester, West Sussex : John Wiley & Sons, Inc., 2017.
Includes bibliographical references and index.
Identifiers: LCCN 2016042844| ISBN 9781119090663 (cloth) | ISBN 9781119090694
(epub).
Subjects: LCSH: Rain forests – Amazon River Region. | Rain forest ecology – Amazon River Region.
Biodiversity – Amazon River Region. Geography – Amazon River Region. |
Plants – Amazon River Region. | Amazon River Region.
Classification: LCC SD160. F67 2017 | DDC 577.340985/44 – dc23 LC record available at
https://lccn.loc.gov/2016042844.

A catalogue record for this book is available from the British Library.

Wiley also publishes its books in a variety of electronic formats. Some content that appears in print may not be available in electronic books.

Cover image: © Sebastián Crespo Photography / Getty Images
Cover Design: Wiley

Set in 10/12pt WarnockPro by SPi Global, Chennai, India
Printed and bound in Malaysia by Vivar Printing Sdn Bhd

1 2017

Dedicated to the memory of my beloved cat, Shaman.

"Das Schöne ist eine Manifestation geheimer Naturgesetze, die uns ohne dessen Erscheinung ewig wären verborgen geblieben." (Beauty is a manifestation of secret natural laws, which otherwise would have been hidden from us forever).

J. W. von Goethe

Source: Goethe, Maximen und Reflexionen. Aphorismen und Aufzeichnungen. Nach den Handschriften des Goethe- und Schiller-Archivs hg. von Max Hecker, Verlag der Goethe-Gesellschaft, Weimar 1907. Aus Kunst und Altertum, 4. Bandes 2. Heft, 1823.

Contents

List of Contributors *xi*
Prologue *xv*

1 **Introduction** *1*
Randall W. Myster
1.1 The Amazon *2*
1.2 The Western Amazon *6*
1.2.1 Case study: Sabalillo Forest Reserve *8*
1.2.1.1 White-sand forest and palm forest plot studies *9*
1.2.1.2 Black-water flooded forest (*igapó*) soil and vegetation studies *10*
1.2.2 Case study: Area de Conservacion Regional Comunal de Tamshiyacu-Tahuayo *11*
1.2.2.1 Plots in *terra firme* forest and black-water flooded forest (*igapó*) *11*
1.2.2.2 Seed predation studies in *terra firme* forest and black-water flooded forest (*igapó*) *13*
1.2.3 Case study: Centro de Investigacion de Jenaro Herrera *13*
1.2.3.1 Soil sampling in various forest types *14*
1.2.3.2 Seed rain sampling in various forest types *15*
1.2.4 Case study: Yasuní experimental station *15*
1.2.4.1 Yasuní *terra firme* forest studies *16*
1.2.4.2 Yasuní white-water flooded forest (*várzea*) studies *17*
1.3 About this book *19*
Acknowledgements *19*
References *19*

2 **A Floristic Assessment of Ecuador's Amazon Tree Flora** *27*
Juan E. Guevara, Hugo Mogollón, Nigel C. A. Pitman, Carlos Ceron, Walter A. Palacios, and David A. Neill
2.1 Introduction *27*
2.2 Methods *28*

2.3 Study area 29
2.3.1 Yasuní 29
2.3.2 Cuyabeno 29
2.4 Herbarium collections 30
2.5 Floristic inventories 30
2.6 Data analysis 31
2.6.1 Estimation of observed and expected tree species richness 32
2.7 Results 32
2.7.1 Observed patterns of tree species richness 32
2.7.2 Estimated number of tree species in Ecuadorian Amazonia 34
2.7.3 Floristic relationships and discontinuities at local and regional scales 36
2.8 Aguarico-Putumayo watershed 37
2.9 Napo-Curaray basin 37
2.10 Pastaza basin region 38
2.11 Cordillera del Cóndor lowlands 39
2.12 What factors drive gradients in alpha and beta diversity in Ecuador Amazon forests? 41
2.12.1 Climate and latitudinal and longitudinal gradients 41
2.13 The role of geomorphology and soils on the patterns of floristic change in Ecuadorian Amazonia 43
2.14 Potential evolutionary processes determining differences in tree alpha and beta diversity in Ecuadorian Amazonia 44
2.15 Future directions 47
References 48

3 Geographical Context of Western Amazonian Forest Use 53
Risto Kalliola and Sanna Mäki
3.1 Introduction 54
3.2 Conditions set by the physical geography 54
3.3 Pre-Colonial human development 57
3.4 Colonial era 59
3.5 Liberation and forming of nations 63
3.6 World market integration and changing political regimes 64
3.7 Characteristics of the present forest use 67
3.8 Present population and regional integration 73
References 77

4 Forest Structure, Fruit Production and Frugivore Communities in *Terra firme* and *Várzea* Forests of the Médio Juruá 85
Joseph E. Hawes and Carlos A. Peres
4.1 Introduction 85
4.2 Methods 88
4.3 Results and discussion 91
4.4 Conclusion 94
References 94

5 **Palm Diversity and Abundance in the Colombian Amazon** *101*
Henrik Balslev, Juan-Carlos Copete, Dennis Pedersen, Rodrigo Bernal, Gloria Galeano, Álvaro Duque, Juan Carlos Berrio, and Mauricio Sanchéz
5.1 Introduction *101*
5.2 Study area *102*
5.3 Methods *103*
5.4 Results *104*
5.4.1 Palms in *terra firme* forests (Figure 5.2) *104*
5.4.2 Palms in floodplain and terrace forests (Figure 5.6) *104*
5.4.2.1 Growth forms *111*
5.4.2.2 Palm architecture *111*
5.4.2.3 Palm species richness (Table 5.1) *113*
5.4.2.4 Palm diversity *113*
5.4.2.5 Palm abundance *113*
5.4.2.6 Palm leaf shape *118*
5.5 Discussion *118*
Acknowledgements *121*
References *121*

6 **Why Rivers Make the Difference: A Review on the Phytogeography of Forested Floodplains in the Amazon Basin** *125*
Florian Wittmann and Ethan Householder
6.1 Introduction *125*
6.2 The geological history of flood-pulsing wetlands in the Amazon Basin *126*
6.2.1 Through the Paleogene *126*
6.2.2 The Miocene *126*
6.2.3 The Quaternary *127*
6.3 Floodplain environments: why rivers make the difference *128*
6.3.1 Trees and flooding *128*
6.3.2 Trees and dispersal in semi-aquatic habitats *130*
6.3.3 Trees and alluvial soils *130*
6.3.4 Trees, hydro-geomorphic disturbance and light regimes *133*
6.3.5 Trees and wetland microclimates *135*
6.4 Conclusions *135*
References *136*

7 **A Diversity of Biogeographies in an Extreme Amazonian Wetland Habitat** *145*
Ethan Householder, John Janovec, Mathias W. Tobler, and Florian Wittmann
7.1 Introduction *145*
7.2 Methods *147*
7.2.1 Habitat description *147*
7.2.2 Vegetation sampling *147*
7.3 Construction of a biogeographic framework *149*

7.4 Results *150*
7.5 Discussion *150*
7.5.1 Insights into local assemblies *152*
7.5.2 Insights into biogeographic processes *153*
7.5.3 Limits of the data *154*
Acknowledgements *154*
References *155*

8 Forest Composition and Spatial Patterns across a Western Amazonian River Basin: The Influence of Plant–Animal Interactions *159*
Varun Swamy
8.1 Introduction *159*
8.2 Methods *162*
8.2.1 Site description and history *162*
8.2.2 Study design *164*
8.3 Analysis *165*
8.3.1 Compositional patterns *165*
8.3.2 Spatial patterns *165*
8.3.2.1 Intra-cohort spatial patterns *165*
8.3.2.2 Inter-cohort spatial patterns *166*
8.4 Results *166*
8.4.1 Compositional patterns *166*
8.4.2 Spatial patterns *171*
8.4.2.1 Intra-cohort spatial patterns *171*
8.4.3 Inter-cohort spatial patterns *172*
8.5 Discussion *173*
References *177*

9 Bird Assemblages in the *Terra Firme* Forest at Yasuní National Park *181*
Andrés Iglesias-Balarezo, Gabriela Toscano-Montero, and Tjitte de Vries
9.1 Introduction *181*
9.2 Methods *182*
9.3 Results and discussion *183*
References *191*

10 Conclusions, Synthesis and Future Directions *195*
Randall W. Myster
10.1 Conclusions *195*
10.2 Synthesis *198*
10.3 Future directions *200*
References *201*

Index *203*

List of Contributors

Balslev, H.
Ecoinformatics & Biodiversity, Bioscience
Aarhus University
Aarhus, Denmark
Email: henrik.balslev@bios.au.dk

Bernal, R.
Instituto de Ciencias Naturales
Universidad Nacional de Colombia
Bogotá, Colombia
Email: rgbernalg@unal.edu.co

Berrio, J. C.
Department of Geography
University of Leicester
Leicester, UK
Email: jcb34@leicester.ac.uk

Ceron, C.
Universidad Central
Escuela de Biología Herbario Alfredo Paredes
Quito, Ecuador
Email: carlosceron57@hotmail.com

Copete, J. C.
Programa de Biología Con Énfasis En Recursos Naturales
Universidad Tecnológica del Chocó
Colombia
Email:
juancarloscopete2010@gmail.com

de Vries, T.
Escuela de Ciencias Biológicas
Facultad de Ciencias Exactas y Naturales
Pontificia Universidad Católica del Ecuador
Quito, Ecuador
Email: tdevries@puce.edu.ec

Duque, A.
Departamento de Ciencias Forestales
Universidad Nacional de Colombia
Email: ajduque09@gmail.com

Galean, G.
Instituto de Ciencias Naturales
Universidad Nacional de Colombia
Bogotá, Colombia
Email: gagaleanog@unal.edu.co

Guevara, J. E.
Department of Integrative Biology
University of California, Berkeley
Berkeley, CA. USA
Email: jeguevara@berkeley.edu

Hawes, J. E.
Animal & Environmental Research Group
Department of Life Sciences
Anglia Ruskin University
Cambridge, UK
Email: joseph.hawes@anglia.ac.uk

Householder, E.
Department of Wetland Ecology
Institute for Geography and Geoecology
Karlsruhe Institute of Technology
Josefstr. 1, Rastatt, Germany
Email: jehouseholder@gmail.com

Iglesias-Balarezo, A.
Rither
Bolivia, Quito
Email: aiglesias9@yahoo.com

Janovec, J.
Facultad de Ciencias Ambientales
Universidad Cientifica del Sur
Lima, Peru
Email: John.Janovec@gmail.com

Kalliola, R.
Department of Geography and Geology
University of Turku
Turku, Finland
Email: risto.kalliola@utu.fi

Mäki, S.
Department of Geography and Geology
University of Turku
Turku, Finland
Email: sanna.maki@utu.fi

Mogollon, H.
Endangered Species Coalition
Silver Springs, CO, USA
Email: hmogollon@stopextinction.org

Myster, R. W.
Biology Department
Oklahoma State University
Oklahoma City, OK, USA
Email: myster@okstate.edu

Neill, D. A.
Universidad Técnica del Norte
Herbario Nacional del Ecuador
Quito, Ecuador
Email: davidneill53@gmail.com

Palacios, W. A.
Universidad Estatal Amazónica
Puyo, Ecuador
Email: walterpalacios326@yahoo.com

Pedersen, D.
Ecoinformatics & Biodiversity, Bioscience
Aarhus, Denmark
Email: dennis.pedersen@bios.au.dk

Peres, C. A.
School of Environmental Sciences
University of East Anglia
Norwich, UK
Email: c.peres@uea.a.cuk

Pitman, N. C. A.
Keller Science Action Center
The Field Museum
Chicago, IL, USA
Email: npitman@fieldmuseum.org

Sanchéz, M.
Departamento de Ciencias Forestales
Universidad Nacional de Colombia
Email: ajduque09@gmail.com

Swamy, V.
San Diego Zoo Institute for Conservation Research
Escondido, California, USA
Email: varunswamy@gmail.com

Tobler, M. W.
San Diego Zoo Global Institute for Conservation Research
Escondido, California USA
Email: mtobler@sandiegozoo.org

Toscano-Montero, G.
Escuela de Ciencias Biológicas
Faculta de Ciencias Exactas y Naturales
Pontificia Universidad Católica del Ecuador
Quito, Ecuador

Wittmann, F.
Department of Wetland Ecology
Institute of Geography and Geoecology
Karlsruhe Institute for Technology
Josefstr. 1, Rastatt, Germany
Email: florian.wittmann@kit.edu
f-wittmann@web.de

Prologue

My first experience in the Amazon occurred in 1995 when I went on a "canned" ecotourism trip to the Rio Napo in eastern Ecuador. Although we saw Anaconda, various monkey species, and a troop of Coatimundi, it was what happened during our return that has stayed with me the longest. Because our plane was departing quite early, we had to leave at four in the morning. We piled into the boat and all was well until we got stuck on a sandbar. At that point, all the men were ordered to disrobe and get into the dark water to push. As I was jumping in, I remembered all the movies and documentaries I had ever seen about the Amazon. I wondered: Would I be attacked from below by a mysterious species unknown to Science, Would the bottom be littered with the corpses of "Indians" murdered by the Conquistadors? or, Would I be swept away, my body melting into a mystical union with the Amazon for all eternity?

I survived to tell the tale and as I worked in the Amazon over the next two decades she dazzled in her beauty, complexity and raw wildness, but not always in the most pleasant way. Perhaps my worst experience was an infected insect bite that landed me in the hospital for two weeks on an IV. During my stay I was told that my leg might have to come off! – but luckily the infection had not reached the bone. Alternatively, I can relate the sense of wonder I felt when, on a clear night, I gazed into the southern sky and saw those stars for the first time, or when I looked into the Yoda-like face of a Uakari monkey from only a few feet away, and waited for it to speak.

And so, the Amazon has been both cruel and deeply satisfying. I have learned to give myself over to her, like riding a horse high in the mountains; trusting her, to take me where I want to go.

R.W.M.

1

Introduction

Randall W. Myster

Abstract

This introduction presents an overview of the key concepts discussed in the subsequent chapters of this book. The book is motivated not just by the Amazon's scientific interest but also by its role in these various ecosystem functions critical to life on Earth. It highlights several of its interactive and higher-order linkages among both abiotic and biotic ecosystem components. The book provides summaries of the author's research in Western Amazonia over the last two decades, in both non-flooded forests and forests flooded with white water and with black water. The Amazonian rainforest is located in the equatorial regions of the South American countries of Brazil, Colombia, Ecuador, Bolivia and Peru. In addition, the Amazonian rainforest influences the entire world's precipitation and weather patterns and, over the longer term, the world's climate. Flooding differs within the Amazon landscape in frequency, timing, duration, water quality, and maximum water depth and height.

Keywords *Amazon landscape; Amazon's scientific interest; biotic ecosystem; black-water floodplain forests; equatorial regions; non-flooded forests; South American countries; world's climate*

The Amazon Basin contains the largest and most diverse tropical rainforest in the world. In particular, the Western Amazon basin is the most pristine and, perhaps, the most complex within the Amazon Basin. This book is motivated not just by the Amazon's scientific interest but also by its role in these various ecosystem functions critical to life on Earth. In this introductory chapter, I first describe that complexity and highlight several of its interactive and higher-order linkages among both abiotic and biotic ecosystem components. Then I include summaries of my own research in Western Amazonia over the last two decades, in both non-flooded forests (e.g. *terra firme*, white sand, palm) and forests flooded with white water (generally referred to as *várzea*) and with black water (generally referred to as *igapó*). Finally, I outline the chapters to come.

Forest structure, function and dynamics in Western Amazonia, First Edition.
Edited by Randall W. Myster.

1.1 The Amazon

When the first Europeans visited the middle of South America in the 1500s, they saw women warriors and named the area after those same figures in Greek Mythology, the Amazons. The two Iberian powers, Spain and Portugal, then fought for control of the Amazon. The Spanish were mainly interested in the wealth of the Incas and so approached the Amazon from the west, but because the Amazon is so flat the Portuguese could colonize (from the east) a much larger part of it. All this was finally resolved by the Treaty of Madrid in 1750, establishing the general boundaries we see today between Brazil (colonized by Portugal) and Bolivia, Peru and Ecuador (colonized by Spain: Hecht 2014).

One may speak of the river itself (the Amazon), the large depression/watershed which surrounds the Amazon and the smaller rivers and streams which drain into it (the Amazon Basin) or the forest that grows in that basin (the Amazonian rainforest). The Amazon river has the greatest discharge of fresh water in the world and is also its second longest river (~6,400 km). It originates in the foothills of the Andean Mountains of South America (Figure 1.1) and runs east into the Atlantic Ocean. It is not constant, however, and has changed over time. For example, climate change during the Pleisotcene (2,588,000 to 11,700 years ago) lead to a drop in sea level which changed its course, and also a rise in sea level that filled its connecting rivers with sediments, creating large floodplains. The Amazon Basin predates the separation of South America from Africa some 110 million years ago (Junk *et al.* 2010) and is generally found below 200 m above sea level (a.s.l.), covering over 8 million km^2 (Hoorn and Wesselingh 2010).

The Amazonian rainforest is located in the equatorial regions of the South American countries of Brazil, Colombia, Ecuador, Bolivia and Peru. Besides the Andes and the Atlantic Ocean, the rainforest is bounded to the north by the Guiana crystalline shield and to the south by the Brazilian crystalline shield

Figure 1.1 Photograph of the Amazon river near where it "officially" begins in Iquitos, Peru.

(Pires and Prance 1985), marked at their edges by cataracts in the rivers and often dominated by grasslands (Myster 2012a). This rainforest is the world's largest tropical rainforest and it is the largest continuous forest of any kind, encompassing over 6 million km^2 (Holdridge 1967; Junk *et al.* 2010; Walter 1973). It produces approximately 20% of the world's oxygen and approximately 10% of the net primary productivity of the entire terrestrial biosphere. Its biodiversity is legendary (present at least since the Pleisotcene), having at least 11,200 tree species (Fabaceae the most common family: Daly and Prance 1989; Hoorn and Wesselingh 2010) and more than 10% of all the species on the planet (Pires and Prance 1985).

Perhaps most importantly for the future of humans, the rainforest interacts intimately with the Earth's carbon (C) cycle, acting both as a carbon "sink", by taking in large amounts of CO_2 through photosynthesis (~15% of the world's total), and as a carbon "source" as, for example, when its plants decay or burn (Rice *et al.* 2004). The Amazonian rainforest contains 20–25% of the world's terrestrial C, with one-third below ground in the soil and two-thirds in the above-ground vegetation (McClain *et al.* 2001). The Amazon rainforest will continue to be a major C player in the future by both contributing to, and suffering the effects of, global warming (Shukla *et al.* 1990).

In addition, the Amazonian rainforest influences the entire world's precipitation and weather patterns and, over the longer term, the world's climate (Keller *et al.* 2004). Evaporation and condensation over Amazonia are engines of global atmospheric circulation (Malhi *et al.* 2008) and this rainforest may even control how much rainfall it itself receives (Pires and Prance 1985). Daily fluctuations in temperature are greater than seasonal fluctuations, that is 27.9°C in the dry season and 25.8°C in the rainy season. Humidity also varies little seasonally, 77% in the dry season and 88% in the rainy season (Prance and Lovejoy 1985) and its prevailing winds come from the east. Because the rainforest is located on the Equator. it has a day length which varies little during the year, as does solar energy input of 767–885 calories per cm^2 per day. Up to 6 m of rain falls every year on the Amazon Basin with a pronounced rainy season that begins in the south. One study showed that 25.6% of that precipitation was intercepted by plants and returned to the atmosphere by evaporation, 45.5% was taken up by plants and transpired, and the rest was absorbed into the soil and/or ran off (through the litter and the soil) into rivers and streams (Salati 1985).

White-water runoff from the Andes appears white because of the high concentrations of dissolved solids, mainly alkali-earth metals and carbonates. This white water has the highest concentration of total electrolytes, phosphorus (P), potassium (K) and other trace elements, of all the Amazonian flood waters, with a pH of around 7 (neutral). Conversely, black water is transparent due to its low amount of suspended matter and has the lowest concentrations of these ions, with high amounts of humic acid (resulting in a pH of between 4 and 5) and clear water is between these two in amounts of sediments and pH. Productivity follows these nutrient trends (Junk and Furch 1985). The resulting Amazonian floodplain forests are the most diverse flooded forests in the world with at least 1,000 tree species, existing since the early Cretaceous (145–66 million years ago: Junk *et al.* 2010). In general, forest in black- and clear-water areas are less diverse, with less litter production, smaller trees and less herbaceous growth than forests in white-water areas (Junk *et al.* 2010). Clear-water rivers come from areas where erosion is less intensive but the more sandy a soil is, the more likely it is to give up its humic substances and create black water. Most leaching occurs during the rainy season and when human activity increases erosion.

The large rainfall results in low fertility soil (and has for millions of years: Jordan 1985), which leads to most of the nutrients being stored in the plant biomass with a fast, efficient closed system of nutrient cycling. P availability mostly limits productivity in *terra firme* forests (McClain *et al.* 2001). The large rainfall also leaches out significant nutrients, such as nitrogen (N), calcium (Ca), P and K out of the leaves and stems of trees and shrubs, adding soluble inorganic and organic substances which allow epiphytic plants to live without a root system. This makes the water cycle a significant interactive link between the soil and the biota. Trees adaptations (e.g. dense root mats at the soil surface) must then be "fine-tuned" (niche-packing) in order to take up and store nutrients efficiently, and this may be one of the reasons for the large biodiversity found in the Amazon. Other reasons may include the relatively constant wet and warm climate (including a predictable flood pulse) and a heterogenous edaphic substrate. Other common adaptations found in Amazon trees include:

- tough, leathery, long-lived leaves;
- supporting colonies of algae and lichens which help recover leached nutrients; and
- sprouting roots from branches and leaves.

Human activities (logging with or without burning and agriculture) can affect many different aspects of the ecosystem – making it another significant interactive linkage – by leading to large losses of biomass (up to 90% C loss compared to the primary forest), with nutrients leaching out of the necromass and into the rivers and streams. Large species diversity may also lead to a large diversity of plant community types, many as yet unknown (Myster 2009, 2012b).

Within the Amazonian rainforest is the dominant, unflooded *terra firme* forest. The *terra firme* has the same physiogomy and many of the same structural characteristics of similar unflooded rainforests throughout the rest of the Neotropics (Everham *et al.* 1996; Kalliola *et al.* 1991; Lopez and Kursar 1999; Whitmore 1989). For example, the Amazonian *terra firme* forest also contains a large amount of above-ground biomass (AGB) and a complex strata of emergent trees, canopy trees, understory trees, palms, understory shrubs, climbing vines, saplings, seedlings, epiphytes, hemi-epiphytic stranglers, lianas, herbs and ferns. Along with this vertical structure, it also has an extensive horizontal structure of various-sized "gaps", light flecks, micro-topological relief and patchy soil nutrient availability in acid, clayey-loamy shallow soils with extensive organic matter in its upper layers. As is true for trees elsewhere in the Neotropics, many of the *terra firme* trees have large buttresses and shallow roots, complex growth/sprouting architectures, and large epiphyte and liana communities (Janzen 1984). Within the broad classification of *terra firme* forest are different types of forest which differ in soil characteristics, for example *terra firme* proper on clay or loam soils, white-sand forests on soils with large amounts of quartz, and palm or swamp forests on standing water (Tuomisto *et al.* 2003).

Flooding within the Amazon Basin generates floodplains and flooded forests, which cover approximately 3 million km^2 (Junk 1989; Parolin *et al.* 2004). Most of this water is the nutrient rich "white" water from the Andes, which creates forests generally called *várzea* (Junk 1984) and the rest is "black" and "clear" water, which is nutrient poor forest runoff and creates forests generally called *igapó* (Junk 1989; Prance 1979; Sioli 1984). There are also forests created from a mixing of the water types (Myster 2009). Differences in nutrient availability in the water may thus be as important in determining the structure, function and dynamics of flooded forests in the Amazon as is the nutrient availability in the soil for unflooded *terra firme* forests. The more studied *várzea* has light levels on the forest floor similar to *terra firme* (1–3% of ambient: Wittmann *et al.* 2010a,b), but flooding creates oxygen deficiency, reduced photosynthesis and low water conductance, so that flooding may be a greater source of mortality than desiccation. In addition, high nutrient levels within these *várzea* forests can lead to trees with rapid growth rates and low wood densities (Parolin *et al.* 2010). Importantly trees within these forests must time their reproduction cycles in relation to the flooding, where some species grow mainly during the flooded times of the year and reproduce when the waters subside, while other species merely "endure" flooding and reproduce only during the dry times of the year (Junk *et al.* 2010). White-water areas are used more than black- or clear-water areas for agriculture – because of more nutrients and the predictability of the flood pulse (Junk *et al.* 2010) – but effects of human land use (Myster 2007a) in white-water areas may be reduced due to flooding (Pinedo-Vasquez *et al.* 2011).

Flooding differs within the Amazon landscape in frequency, timing, duration, water quality, and maximum water depth and height (Junk *et al.* 2010; Myster 2009). Such variation within the flooding gradient (Myster 2001) greatly affects the distribution and abundance of plant species (Ferreira and Stohlgren 1999; Junk 1989; Lamotte 1990), leading to inundated vegetative associations created by the rise of the water table on a regular, seasonal basis. In general, flooded forest types, vegetation formations and plant communities lie on a continuum defined by:

- the duration of the aquatic and terrestrial phases of the annual cycles; and
- the physical stability of the habitat influenced by sedimentation and erosion processes (Junk *et al.* 2010).

In general, the soils in the floodplain are less acid than those of the *terra firme*, but may have a greater concentration of exchangeable cations such as Mg^{+2} and Na^{+2} (reviewed in Honorio 2006).

Because the Amazon and its tributaries are very dynamic – often changing their routes within a time span of a few decades (Junk 1989; Kalliola *et al.* 1991; Pires and Prance 1985) – it may well be that forests that are unflooded today were flooded in the past and *vice versa*. It is not surprising that many *terra firme* species establish ecotypes (Myster and Fetcher 2005) in the flooded forest (Wittmann *et al.* 2004; 2010a,b). For example:

- the *terra firme* species *Guazuma ulmifolia* and *Spondias lutea* have developed flood-resistant ecotypes now found in *várzea*;
- *várzea* species such as *Ceiba pentandra* and *Pseudobombax munguba* occur in *terra firme*; and
- several species of the genus *Maquira* occur in both unflooded and flooded forests.

This ecotropic dynamic and the high species richness of the surrounding Amazonian *terra firme* rainforest, which disperse seeds into flooded forests – when combined with flooding and its associated environmental heterogeneity – suggests that flooded forests will have a unique biology and ecology (Kalliola *et al.* 1991). Furthermore, it is expected that flooding creates specific tree species zonational distributions (Whitmann *et al.* 2010a), largely determined by the submergence tolerance of their seedlings (Parolin *et al.* 2004). Finally the predictability of the flood "pulse" (Junk *et al.* 2010) – both past and present – facilitates adaptation and thus, along with differences in the surrounding biota and a variety of soil types (Honorio 2006; Junk 1989), may help create complex and diverse forest associations throughout the Amazon Basin (Myster 2009).

1.2 The Western Amazon

Studies of tree endemism in the Amazon (ter Steege *et al.* 2006; Whitmann *et al.* 2013) show that there is a natural division between the Western Amazon rainforest (Myster 2009) and the Central Amazon rainforest (Junk *et al.* 2010), where the Japurá river joins the Amazon river (Figure 1.2). In this book, I will accept that evidence and define Western Amazonia as everything in the Amazon Basin west of latitude 65°W, including the Mamiraua Reserve in Tefé, Brazil.

This western part of the Amazon Basin is composed of young and relatively fertile sediment, which makes the flora of more recent origin than that of Eastern Amazonia (reviewed in Dumont *et al.* 1990; Terborgh and Andresen 1998). Furthermore, the higher fertility of these soils may lead to more forest turnover and evolution, resulting in the high biodiversity sampled here compared to Central Amazonia (Bierregaard *et al.* 2001; Gentry 1993). Indeed the Western Amazon rainforest contains some of the most diverse areas or "hotspots" on Earth (Myers *et al.* 2000; Myster 2007b) due, in part, because it is largely intact, pristine and unaffected by human activity (<5% of its total area is in secondary forest: Gorchov *et al.* 2004; Neff *et al.* 2006; Soares-Filho *et al.* 2006) and so suffers the smallest loss of biodiversity and forest fragmentation (Bierregaard *et al.* 2001) in all of the Amazon. The large biodiversity of the Western Amazon may also be a consequence of the large areas within it, which have been surrounded by only rainforest for millions of years, without border effects or edge effects. Because the Western Amazon rainforest has extensive variation in edaphic, climatic, topographic and geological conditions, a large number of distinct plant communities are expected (Myster 2009, 2012b). Indeed, sampling has shown

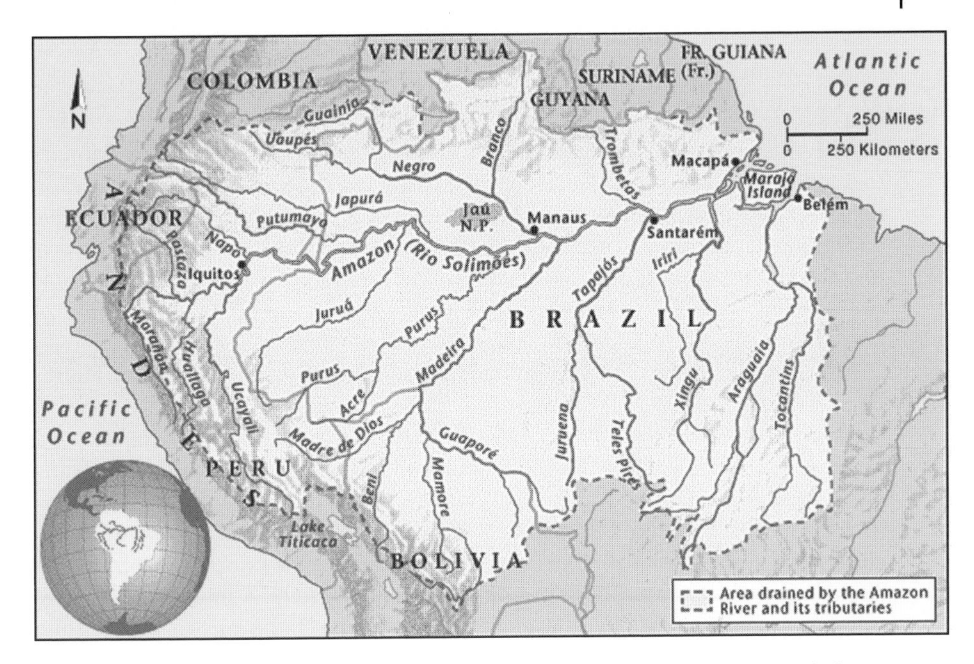

Figure 1.2 Map of the Amazon river and some of the other rivers that drain into it, including the Japurá river, with an outline of the Amazon Basin.

that up to 59% of recorded species are found in only one plot among the many sampled (Duivenvoorden 1995), and shared species between wetland and upland forests is very low (Dumont *et al.* 1990).

All tropical rainforests are hot and humid with large biodiversity. However, in addition to the global and ecosystem issues raised before, the Amazon in general, and the Western Amazon in particular, are also unique because of their high level of complexity – the many ways that different components of the Amazon ecosystem interact – and often at a higher-order than other tropical forests. Some of those interactions are outlined above, and in coming chapters authors will discuss many more Amazonian interactions within their own area of interest. I also point to these recent studies that have highlighted important Amazon ecosystem interactions among:

1. soil type, land use and tree growth (Moran *et al.* 2000);
2. water availability and forest biodiversity (Paine *et al.* 2009);
3. deforestation, fire and drought (Davidson *et al.* 2010);
4. recruitment and scatter-hoarding (Russo *et al.* 2005);
5. recruitment and seed dispersal by fish (Anderson *et al.* 2009);
6. hunting of large primates, richness and density (Nunez-Iturri and Howe 2007);
7. soils, geography distances and floristics (Ruokolainen *et al.* 2007);
8. flooding duration and the availability of phosphorus, iron and aluminum (Chacon *et al.* 2008);
9. soils, herbivory and plant defenses (Fine *et al.* 2004);

10. forest responses, drought and the C cycle (Phillips *et al.* 2009; Saleska *et al.* 2007); and
11. climate change and flooding patterns (Langerwisch *et al.* 2012).

With this ecological background in mind, and as background to the chapters that follow, it may be useful to review the ideas and different approaches used in the past to investigate Neotropical forests. First, it has certainly been a positive development over the last few decades of Neotropical forest research that several permanent forest plots has been set up and sampled on a regular basis for a considerable time. These have included large – usually 25–50 ha – plots in *terra firme* forest (i.e. Barro Colorado Island in Panama, Yasuní National Park (YNP) in Ecuador, La Planada in Colombia, Luquillo Experimental Forest in Puerto Rico: Brokaw *et al.* 2012) and smaller plots/transects in various Neotropical forests (Balslev *et al.* 1987; Myster 2007b, 2010; Parolin *et al.* 2004; Worbes *et al.* 1992; author unpub. data). The data collected has advanced our understanding of the structure of these forests and how that structure changes through time (Brokaw *et al.* 2012; Junk *et al.* 2010; Losos and Leigh 2004). I applaud those efforts and hope they will continue. In addition to plot sampling, approaches have included tabulation of common, or otherwise defined important, species traits (ecophysiological: Parolin *et al.* 2004; Poorter 1999; phenology: Parolin *et al.* 2010; mycorrhizal association: Falster *et al.* 2015; Meyer *et al.* 2010; Myster *et al.* 2013), measurement and estimation of different aspects of the C cycle (Baker *et al.* 2007; Schongart *et al.* 2010; Townsend-Small *et al.* 2005) and other biogeochemical cycles (Calle-Rendon *et al.* 2011; Kern *et al.* 2010), remote sensing used to discover large-scale vegetation patterns and tree associations (Melack and Hess 2010; Peixoto *et al.* 2009) and seed/seedling mechanistic experiments (e.g. Kubitzki and Ziburski 1994; Paine and Beck 2007; Wittmann *et al.* 2010b).

While these plots and other research approaches have had success, field experimentation and modeling approaches tried in Neotropic forests over the last 30–40 years (Carson and Schnitzer 2008) continue to be a challenge, especially for those trying to discover what controls forest dynamics: in particular, those approaches that include hypotheses about static, unchanging patterns of biodiversity (Connell 1971; Janzen 1970), which then need to be "maintained" over time (McDade *et al.* 1994; Zimmerman *et al.* 2008), and co-existence of species (Dalling and John 2008; Mabberley 1992). Alternatively, the authors in this book will progress with better methods of investigation into forest structure, function and dynamics, which I will put together into a new and better synthesis of Neotropic rainforest investigation, in the concluding chapter.

1.2.1 Case study: Sabalillo Forest Reserve

I now review my own studies in the Western Amazon. My first study site was Sabalillo Forest Reserve (SFR: 3°20′3″S, 72°18′6″ W: Frederickson *et. al.* 2005; Moreau 2008) established in 2000 and operated by Project Amazonas. SFR is located on both sides of the upper Rio Apayacuo, 172 km east of Iquitos, Peru. The reserve is part of 25,000 acres set aside over the last decade and is comprised of low, seasonally inundated river basins of the upper Amazon. The substrate of these forests is composed of alluvial and fluvial Holocene sediments from the eastern slopes of the Andes. Annual precipitation has been measured as 3,297 mm per year (Choo *et. al.* 2007). Within the SFR are areas of black-water runoff, which create *igapó* forests of differing frequency, duration and maximum water column height, where the rainy season is between November and April. In addition, *terra firme* forest is common as well as white-sand forest and palm forest (which exists on standing water: Holdridge 1967). White-sand forests form on white-sand soils of extreme infertility, made up mainly of quartz podzols. The FSR white-sand forest is called "varillal", which is defined as a dense forest with straight and thin trees between 10 and 20 m high (Honorio 2006; Myster 2009). White-sand forest has a species diversity between *terra firme* forest (which is higher) and palm forest (which is lower: Duivenvoorden *et al.* 2001), and common families include Leguminosae, Clusiaceae, Malvaceae, Euphorbiaceae and Icacinaceae (Honorio 2006).

1.2.1.1 White-sand forest and palm forest plot studies

In my first investigation at SFR in June 2013, my two student field assistants and I set up a 1-ha plot (200 m × 50 m) in a white-sand forest and in a palm forest at SFR. We tagged and measured the diameter at breast height (dbh) of all trees at least 10 cm dbh in a hundred 10 m × 10 m continuous subplots using the protocol as for the 1-ha plot at Yasuní and subsequent plots. Plots of this size have been used to study forests in the Amazon for decades. The tagged trees were identified to species, or to genus in a few cases, using Romoleroux *et al.* (1997) and Gentry (1993) as taxonomic sources. We also consulted the Universidad Nacional de la Amazonia Peruana herbarium and the website of the Missouri Botanical Garden (www.mobot.org).

In the white-sand forest I found:

1. there were a total of 15 families and Clusiaceae was by far the most common family, which also had the most genera and the most species;
2. Malvaceaea, Myrtaceae and Rubiaceae were also common, and there were 2 families with only 1 stem;
3. most families had a monotonic decline in stem number as stems became thicker and the species *Pachira brevipes* and *Caraipa tereticaulis* were the most common;
4. stems conformed to a "reverse J" size distribution pattern for all stems, as was true of *terra firme* samplings;
5. the average stem dbh was slightly less than *terra firme* plots, but basal area and AGB was greatly reduced compared to *terra firme* plots; and
6. the forest canopy was 14% closed and the trees were randomly dispersed.

In the palm forest plot I found:

1. there were a total of 6 families and Arecaceae was by far the most common family, which also had the most genera and the most species, but Clusiaceae, Fabaceae and Lecythidaceae were also common;
2. all families had a monotonic decline in stem number as stems became thicker and the species *Socratea exorrhiza* and *Lepidocaryum tenue* (both palms) were most common;

3. total number of stems was a bit more than white-sand forest, but much less than each of the three *terra firme* plots;
4. stem size distribution was comparable to white-sand forest, but average stem size was much reduced when compared to both white-sand and one *terra firme* sampling;
5. whereas species richness was much less for palm as well (compared to the other forests), the reduction in basal area and AGB was smaller; and
6. palm was a dimmer, denser forest than white-sand and showed no clumping.

I conclude that:

1. white-sand forest is less complex in floristics and various aspects of structure compared to *terra firme* forests, which may be due to its soils being of low fertility, but similar in number of families and species to some black-water *igapó* forests that are under water at least five months of the year, and
2. palm forest is less complex in floristics compared to both white-sand and *terra firme* but was mixed for other structure characteristics, which may be due to its soils being water-logged and lacking in oxygen for roots (author, unpublished data).

1.2.1.2 Black-water flooded forest (*igapó*) soil and vegetation studies

I next investigated the floristics, and how soil bulk density affects physical structure, of successional *igapó* forests at SFR. In June 2013, my field assistants and I sampled five successional areas – island, oxbow lake, river margin, sandy beach, side creek – close to a black-water river at SFR. In each area we set up 105 m × 5 m continuous plots for a total sampling area of 250 m^2 per area. We measured the dbh of each tree at least 1 cm dbh within each plot. The dbh measurement was again taken at the nearest lower point where the stem was cylindrical and for buttressed trees it was taken above the buttresses. We identified its species, or to genus in a few cases, using Romoleroux *et al.* (1997) and Gentry (1993) as taxonomic sources. We also consulted the Universidad Nacional de la Amazonia Peruana herbarium and the website of the Missouri Botanical Garden (www.mobot.org).

In addition, we collected three soil samples in each area by driving a 3-inch diameter ring into a depth of 10 cm and extracting the soil. Back in Iquitos, each soil sample was dried for three 4-minute cycles in a microwave and then weighed. Soil bulk density was computed as weighed soil sample/volume of container, expressed as g/cm^3. I found (author, unpublished data):

1. a total of 24 plant families where Urticaceaea was the most family present, but Rubiaceae and Euphorbiaceae were also common. Most families had a monotonic decline in stem number as stems became thicker and there were no stems with a dbh greater than 29 cm. The most common species were *Cecropia membranaceae, Sapium glandulosum, Pourouma guianensis* and *Byrsonima arthropoda.*
2. soil bulk density was highest in the sandy beach and lowest in the forest under water for the longest duration. The greatest number of stems was 47 in the island area and 18 in the forest under water for the longest duration.

Mean stem size, species richness, Fishers α, basal area and AGB were lowest in the sandy beach and highest in the forest under water for the shortest duration; and

3. linear regression analysis showed that soil bulk density could best predict mean stem size, species richness and Fishers α. Results showed that as soil became less sandy and with more clay content, all structural parameters except stem number increased, suggesting an increase in forest community complexity as soil increases in water retention capacity and nutrients.

I conclude that the severe and unpredictable flooding seen in these successional areas reduces forest structure more than the predictable and seasonal flooding that black-water floodplain forests receive.

1.2.2 Case study: Area de Conservacion Regional Comunal de Tamshiyacu-Tahuayo

My second study site in the Western Amazon was the Area de Conservacion Regional Comunal de Tamshiyacu-Tahuayo (ACRCTT: www.perujungle.com, Gottdenker and Bodmer 1998; Myster 2007b, 2009, 2010) located in Loreto Province, 80 miles southeast of Iquitos, Peru (~2°S, 75°W) with an elevation of 106 m. The reserve is part of one of the largest (270,654 ha) protected areas in the Amazon, containing wet lowland tropical rainforest of high diversity (Daly and Prance 1989). It is comprised of the low, seasonally inundated river basins of the upper Amazon and named after two of the major white-water rivers (the Tahuayo and the Tamshiyacu), which form boundaries to the north and west and create large fringing floodplains (Junk 1984). The substrate of these forests is composed of alluvial and fluvial Holocene sediments from the eastern slopes of the Andes. Annual precipitation ranges from 2,400–3,000 mm per year, and the average temperature is relatively steady at 26°C.

The rainforest itself is divided into distinct communities defined by their flooding regime where the rainy season is between November and April (Kalliola *et al.* 1991), for example non-flooded *terra firme*, *igapó* forest under water 1–3 months per year and *igapó* forest under water 4–6 months per year. Common tree genera in high restinga are Chorisia, Eschweilera, Hura, Spondias and Virola. Low restinga and tahuampa contain tree species such as *Calycophyllum spruceanum*, *Ceiba samauma*, *Inga* spp., *Cedrela odorata*, *Copaifera reticulata*, *Phytelephas macrocarpa*, with under-story palms such as *Guazuma rosea* and *Piptadenia pteroclada* (Daly and Prance 1989; Myster 2007b; Prance 1979; Puhakka *et al.* 1992).

1.2.2.1 Plots in *terra firme* forest and black-water flooded forest (*igapó*)

Within each of three forest types which differed in flood duration and maximum water depth (unflooded *terra firme*, *igapó* forest under water 1–3 months per year, *igapó* forest under water 4–6 months per year), five areas were selected in June of 1997. These were primary, unlogged forest and had fresh average-sized (100–300 m^2: Brokaw 1982) gaps within them. I sampled all tree stems at least 10 cm in dbh within each of 100 m^2 (10 m × 10 m) microsites. This makes a total of 30 microsites, the three forest types with their gaps and five reps (6 × 5). This data is housed in the archives of the LTER site in Puerto Rico (LTERDBAS #150) and may be accessed through their website (luq.lternet.edu).

I found (Myster 2007b):

1. common species existed between the two *igapó* forests and their gaps and between the gaps of the two *igapó* forests;
2. tree richness was maximum in *terra firme* forests, medium in the least flooded *igapo* and their gaps, and smallest in the most flooded *igapo* gaps;
3. there were less stems in gaps compared to forests and less stems in forests as flooding increased, except again in the least flooded *igapo* gaps; and
4. dominance-diversity curves have more dominance by single species in the *terra firme* gap plots compared to other gaps and in *terra firme* forest compared to other forests.

In general, while some aspects of structure such as tree stem density is largely determined by treefall gap dynamics, tree composition is determined by the flooding regime. Finally, a jump in tree richness in wet forests and wet gaps compared to other plots suggests a "mass effects" hypothesis where species from dry and very wet forest and gaps have overlapping ranges in the wet forest and gap. This effect may help explain the high species diversity seen in this part of the Amazon.

Further analysis of this dataset (Myster 2010) showed a significant effect of treefall gap formation (gap, no gap) on canopy average height, canopy maximum height, basal area, density, AGB, turnover and alpha diversity, and a significant effect of forest type on species richness, genera richness, density, turnover and alpha diversity. In general, there were fewer trees, but they were larger, and more productive in the forest plots compared with the gap plots; and the most flooded plots had fewer trees, species and genera compared with both the less flooded forest and non-flooded forest. Also the greatest amount of turnover was found in the most flooded forests, and the intermediately flooded forest had the greatest richness and alpha diversity. Canopy structure was determined by traditional gap dynamics, but much of canopy diversity depended on the type of forest, tree density decreased as flooding increased, especially among the smallest stems, and there was evidence to suggest that the high biodiversity of the Amazon may be maintained in part by the existence of moderately flooded forest and gaps.

In the third in this series of papers from the same sampling (Myster 2015b), I found:

1. trees became less dense as flooding increased, more than due to tree fall;
2. trees were dispersed uniformly in the forest and clumped in the gaps for the least flooded forest, but became random as flooding increased;
3. canopy coverage followed the same trends as stem density;
4. tree stem size distribution showed smaller stems in gaps compared to adjacent forest for all forest types, where gaps were more consistent in maintaining this pattern than forests as flooding increased; and
5. canopy cover was dominated by smaller stems in gaps compared to forests and forests lost some of their smaller stems – with their canopy contribution – as flooding increased.

I conclude that flooding placed a greater stress on these forests than tree fall, where Amazonian forests may be an ecosystem where gradients and disturbances overlap in their traditional roles and present plants with similar cues.

Finally, I set up a 1-ha plot in the most flooded *igapó* forest in 2010 and sampled it in the same way as the previous 1-ha plots (Myster 2013). I found that stems conformed to the reverse J-size pattern, total stems were lower than other *igapó* forest ranges with a slightly larger average dbh, trees were clumped at a higher degree than that found in the 1-ha plot I sampled in *várzea* forest with 12% canopy closure, while the basal area and AGB was less than both other *igapó* samplings and the 1-ha *várzea* study plot, and flooding produced reduced basal area in *igapó*, and smaller stems, stem densities and AGB for both flooded forests. I conclude that in both study plots, there was a reduction of tree stem density and structure (basal area, AGB) with flooding, which reduced even more as months under water increased. More sampling in these forests is needed, however, before a conclusion about which aspect of the flooding regime – for example, water quality, flooding duration or frequency – is most important in determining different aspects of forest structure.

1.2.2.2 Seed predation studies in *terra firme* forest and black-water flooded forest (*igapó*)

I set out seeds on transects for a week in seven different tropical forests. I found:

1. seed predation took more seeds then either seed pathogenic disease or germination for most seed species, but there were a few species that lost more seeds to pathogens than predators. Germination had the lowest percentage (%) for most species, but again there were a few species where germination % was higher than pathogens;
2. within the unflooded forests, there was the most predation in *terra firme* forest, palm forest (at SFR) lost more seeds to pathogens than predators, and white-sand forest (at SFR) predation levels were between the other two forest types;
3. within the flooded forests, predation decreased as water went from white to black (at the same inundation levels) and predation decreased monotonically (and pathogens increased monotonically) as months under water increased in black-water forests; and
4. there was significantly more seed predation in the unflooded forests compared to the flooded forests, but significantly more germination in the flooded forests compared to the unflooded forests.

I conclude that seed predation is the major post-dispersal filter for regeneration in these forests, but pathogenic disease can play a major role, especially in forests that have water in them for long periods each year, so that flooding may change those forests dramatically by altering the actions of seed mechanisms and tolerances (author, unpublished data).

1.2.3 Case study: Centro de Investigacion de Jenaro Herrera

My third study site in the Western Amazon was the biological station Centro de Investigacion de Jenaro Herrera (CIJH) operated by the Instituto de

Investigaciónes de la Amazónia Peruana (www.iiap.org) and located 2.5 km from the town of Jenaro Herrera, 200 km south of Iquitos on the east margin of the Ucayali river in the Province of Requena, Department of Loreto, Peru (4°54′S, 73°40′W: Honorio 2006). The mean annual temperature is 26.0°C with a range between 25.1°C and 26.5°C. The mean annual rainfall is 2,724 mm with two dry seasons, the more severe between June and September and the less severe between December and March. Here the non-flooded *terra firme* forests have soils that are highly weathered, acidic, nutrient-poor and clayey-loamy (oxisols: Spichiger *et al.* 1996).

Dominating the non-flooded portions of the study site are low terrace broad leaf forest (*terra firme*-low) and high terrace broad leaf forest (*terra firme*-high) with scattered patches of white-sand-varillal (with dense growth and trees up to 20 m tall), white-sand-chamizal (with shrubs and scattered trees less than 8 m tall), and low terrace palm (dominated by palms such as *Mauritia flexuosa, Oenocarpus bataua, Euterpe precatoria* and *S. exorrhiza*). Common genera in the high and low terrace broad leaf *terra firme* forests include *Eshweilera, Pouteria, Oenocarpus, Miconia* and *Protium*, in white sand-varillal *Pachira, Haploclathra* and *Macrolobium*, and in white-sand-chamizal *Caraipa, Pachira, Macrolobium, Calophyyum, Haploclathra* and *Platycarpum* (Honorio 2006). *Terra firme*-high terrace forest is located adjacent to rivers (Pires and Prance 1985), but is rarely flooded. *Terra firme*-low terrace is not located on ridges adjacent to rivers and has clay soils. Palm swamps (Montufar and Pintaud 2006) occur in depressions or low-lying patches with poor drainage. White-sand soils consist mainly of quartz with individual trees having slender boles and roots concentrated at the soil surface (Klinge *et al.* 1990). These forests are divided into varillal (dense with straight and thin trees between 10 m and 20 m high) and chamizal (shrubs 3 m high and scattered trees more than 8 m high: Honorio 2006).

1.2.3.1 Soil sampling in various forest types

To better understand Amazon soils and how they differ between non-flooded and flooded forests, I took soil samples (author, unpub. data) in eight different forests at two locations in the Peruvian Amazon. In May 2009, I established transects in each of the eight study forests found at CIJH (*terra firme*-low, *terra firme*-high, white-sand-varillal, white-sand-chamizal, palm) and at ACRCTT located also in Peru and discussed in Section 1.4 (high restinga, low restinga, tahaumpa). Along each transect at microsites of approximately 10-m intervals, I took five soil samples without litter. Data was used for one-way analysis of variance tests for each soil parameters between the flooded forests (pooled for all three types) and the most common non-flooded forests, *terra firme*-low and *terra firme*-high. I found:

1. soil pH of the non-flooded forests was very similar to flooded forests, but became more basic with increased flooding;
2. soil organic matter was lowest in the two non-flooded *terra firme* forests and also increased with flooding;
3. N was lowest in the palm forest, P was lowest in *terra firme*-low terrace forest and K was lowest in the *terra firme*-high terrace forest;

4. while N decreased sharply with flooding, both P and K increased with length of the flooding period; and
5. for some non-flooded forests there was a correspondence between soil fertility and floristic similarity.

I conclude that flooding has significant effects on nutrient availability of Amazonian forest soils, by increasing the concentration of some nutrients but decreasing it for others.

1.2.3.2 Seed rain sampling in various forest types

Because of the importance of seed processes in determining rainforest regeneration and dynamics, I sampled the seed rain (Myster 2015c) in six different tropical forests across the Amazonian landscape. I set up seed traps and took seed samples in three unflooded forests at CIJH (*terra firme*, white-sand-varillal, palm) and three flooded forests at ACRCTT (high restinga, low restinga, tahaumpa) in Peru over a period of one year. I found:

1. all forests except tahuampa contained seeds of tree species that have been sampled in other studies within a forest of the same type, but in all forests there were seeds of several tree species that have not yet been sampled within their forest type;
2. total seed load peaked in the early part of the year – near the end of the rainy season – and then decreased monotonically over the remainder of the year for all forests;
3. species richness was greater in unflooded forests compared to flooded forests and the largest number of species were found in *terra firme*;
4. seeds were more evenly distributed among species in the unflooded forests compared to the flooded forests; and
5. alpha diversity was much greater in *terra firme* compared to all other forests.

I conclude for the unflooded forests that seed species number and richness increased with soil fertility, but for the flooded forests seed species number and richness decreased with months under water. When taken together, results suggest that for forests across the Amazonian landscape, differences in flooding regime may have a greater effect on both seed rain load and seed species richness than differences in availability of soil nutrients.

1.2.4 Case study: Yasuní experimental station

My fourth and last study site was the Yasuni Experimental Station (0°41′S, 76°24′W), operated by the Pontificia Universidad Catolica of Ecuador and located within the YNP of eastern Ecuador (Duivenvoorder *et al.* 2001; Metz *et al.* 2008; Myster 2012c; Myster and Santacruz 2005; Svenning 1999). Most of the YNP is *terra firme* forest, classified as lowland tropical rainforest (Holdridge, 1967). The mean annual rainfall is 3,081 mm, with the wettest months April to May and October to November. August is the driest month and the mean monthly temperature varies between 22°C and 35°C. Soils in the park have been described as clayey, low in most cations but rich in aluminium and iron, whereas soils at the station in *terra firme* forest are acidic and rich in exchangeable bases with a texture dominated by silt (Tuomisto *et al.* 2003). The park has low topographic variation with a mean elevation of approximately 200 m a.s.l.

The station is also the site of a long-term 50-ha vegetation plot in *terra firme* forest, maintained by the Smithsonian Tropical Research Institute (Losos and Leigh 2004), parts of which have been sampled with summaries of species found, densities and aspects of forest structure reported (Valencia *et al.* 2004a,b,c, 2009). Also found is floodplain *várzea* forest – located next to the nutrient-rich white-water Tiputini river – which is underwater off and on between the months of October and April for a few weeks to a maximum depth of 3 m.

1.2.4.1 Yasuní *terra firme* forest studies

My research in *terra firme* forest at Yasuní began by investigating how tree seedlings respond along a 100-m transect set up 10 m outside and parallel to the 50-ha plot (Myster 2012c), in order to better understand availability of plant resources on the forest floor in the Amazon, and also to show the effect of their heterogeneity on tree seedlings. I first described the spatial and temporal variation of light and soil water along a 100-m transect, 10 m outside the southern border of the 50-ha plot, in *terra firme* forest for 6 months, and then recorded seedling responses of 3 common tree species planted on that transect after 1 years' growth. I found that:

1. the spatial heterogeneity across the transect was greater than the temporal heterogeneity at any given microsite on the transect for both light and water and there was a positive correlation between them;
2. *Couepia obovata*, the largest seeded and the only subcanopy tree, survived best and showed both the largest relative height growth rate (RHGR: Falster *et al.* 2015) and the largest specific leaf area (SLA), while among the two early successional trees, *Tapirira guianensis* had the largest leaf area ratio (LAR) and the largest leaf mass ratio (LMR) and *Duguethia spixiana* had the largest root/shoot ratio (RTOS);
3. for *T. guianensis*, SLA increased with increasing light and soil water potential predicted both increasing LMR and decreasing RTOS with increasing soil water; and
4. soil water potential could also predict increasing LAR with increasing water for *D. spixiana* and, for *C. obovata*, soil water potential predicted more survivorship, LMR and RHGR but less RTOS, all with increasing soil water.

I conclude that some sub-canopy trees may survive and grow more than open-canopy trees when presented with water stress in the forest understory, and that within the ranges of light and soil water sampled here, plants responded more to spatial variation in water compared to light.

My research continued by investigating regeneration (Grubb 1977) in closed *terra firme* forest and within its gaps by setting out seed traps and collecting soil samples in those microsites (Myster 2014). I also set out seeds and seedlings in the same forest and gaps and later collected them scoring various survivorship and growth parameters. I found that:

1. total number and richness of dispersed seeds was greatest in the *terra firme* gaps compared to *terra firme* forest;

2. *terra firme* forest had less seedlings germinating from soil samples than *terra firme* gaps;
3. on average, 30% of seeds remained after two weeks in the field and species seed losses were significantly different in both forest and gap;
4. more large seeds than small seeds were lost to pathogens and those losses were greater in the *terra firme* forest than *terra firme* gap;
5. as seed mass increased, seedlings survival rates also increased but growth rates declined in both forest and gap; and
6. more seedlings survived in gaps compared to closed forests, but they grew faster in *terra firme* forest with a larger LAR.

My last research project in *terra firme* forest was a sampling of insects (Myster and Santacruz 2005) as possible agents of the seed predation just reported. We sampled in the same closed *terra firme* forest and *terra firme* gaps, where seed dishes where placed. In the closed *terra firme* forest, we found arthropods in the orders Diptera (92 individuals), Hymenoptera (60), Collembola (5), Orthoptera family Grillidea (5), Coleoptera (1), Acaros (1) and Aracnida (1). In the *terra firme* gaps, we found insects in the orders Diptera (80 individuals), Hymenoptera (52), Collembola (21), Coleoptera family Staphilinidae (4), Orthoptera family Grillidea (2) and Hemiptera family Cydhidea (1). Number of individuals and orders was similar between these two areas. The rankings of these orders became different only among those with the smallest abundances.

1.2.4.2 Yasuní white-water flooded forest (*várzea*) studies

My *várzea* research began in May and June of 2010, when I set up a 1-ha plot next to the Tipitini river (a tributary of the Napo river) and also located a few hundred meters from the 50 ha plot. The 1-ha plot was subdivided into 100 continuous 10 m × 10 m sublots. In each subplot I and my field assistants tagged, identified and measured the dbh of all trees at least 10 cm dbh, expanding on past sampling of Amazon flooded forests (Balslev *et al.* 1987; Myster 2007b, 2010; Parolin *et al.* 2004; Worbes *et al.* 1992). The dbh measurement was taken at the nearest lower point where the stem was cylindrical and for buttressed trees it was taken above the buttresses. The data are archived at the Luquillo Experimental Forest, Puerto Rico as LTERDBAS#172, part of the LTER program funded by the US National Science Foundation. One may visit their website (http://luq.lternet.edu) for further details.

I found (Myster 2013, 2015a) in the 1-ha plot that:

1. the seven most common families sampled were also among the top ten families sampled in the 50-ha plot, but most of the rare families were not;
2. at the genus and species taxonomic level, similarities with the 50-ha plot disappeared, except for the genera Cecropia, Lachornea, Inga, Zygia, Eschweilera and Virola and the species *Iriartea deltoidea* and *Coccoloba densifrons;*
3. the 1-ha plot lost stems with flooding but that loss was mainly in the smaller size classes, leading to a proportionally greater number of larger trees than the *terra firme* forest and a larger basal area for stems at least 40 cm dbh; and

4. because the flooded forest loses families, genera and species proportionally more than they lose stems, Fisher's α was lower in the flooded forest compared to *terra firme* forest.

In addition, stems conformed to a reverse J-size pattern for stems less than 40 cm dbh, trees were clumped at a low level with 45% canopy closure, basal area was within other *várzea* forest ranges, and above-ground biomass was lower, and flooding produced smaller stems, stem densities and above-ground biomass compared to the *terra firme* forest sampled as part of the 50-ha plot.

Next I investigated (Myster 2014) on how *várzea* forest regeneration is affected by treefall gap formation, by setting out seed traps and collecting soil samples in the forest and its gaps. I also set out seeds and seedlings in the same forests and gaps and later collected them, scoring various survivorship and growth parameters. I found that:

1. the total number and richness of dispersed seeds was greater in the *várzea* gaps compared to the *várzea* forest;
2. likewise, a greater number of seedlings germinated from *várzea* gap soils compared to the *várzea* fores;
3. on average, 30% of seeds remained after two weeks in the field and species seed losses were significantly different;
4. more large seeds than small seeds were lost to pathogens, and those losses were greater in the *várzea* forest compared to the *várzea* gap microsites;
5. as seed mass increased seedlings survival rates also increased but growth rates declined; and
6. more seedlings survived in *várzea* gaps compared to closed *várzea* forests, where they grew faster with a larger LAR.

When we combine these regeneration results with those in *terra firme* forests (discussed eariler), we see that most regeneration mechanisms had the greatest seed and seedlings losses and the slowest growth rates in intact *terra firme* forest, medium losses and growth rates in *terra firme* gaps and intact *várzea* forest, and the smallest losses and fastest growth in *várzea* gaps. These results are consistent with viewing flooding as a disturbance, like gap formation, both of which structure the Amazon rainforest (Myster 2007, 2010, 2015).

Finally, I compared *igapó* sampling at ACRCTT and *várzea* sampling at YRS (Myster 2016). There was species variation among the plots within both forest types, but little variation in physical structure. The *igapó* plot had 16 families, 29 genera and 31 species, with Fabaceae the most common family which also had the most genera and the most species. The *várzea* plot had 42 families, 91 genera and 159 species with Fabaceae again the most common family, which also had the most genera and the most species. There were only four species in common. In general, the *várzea* plot had more stems, and more large stems (at least 40 cm dbh) than the *igapó* plot, but mean stem size was very similar. Structural comparison to *terra firme* 1-ha plots showed it had more stems, thicker stems and more above-ground biomass compared to either of these pooled 1-ha flooded plots. Finally, all study plots conformed to the reverse J-stem size distribution pattern for all stems.

1.3 About this book

Here I take advantage of my many years working in Ecuador and Peru, and my three awards from the Fulbright Foundation, to edit a book based solely on the Western Amazon which has never been done. While all tropical rainforests are hot, humid places with large biodiversity, the Amazonian rainforest in general and the Western Amazonian rainforest in particular are different and important. They are places of more complex linkages and higher-order interactions among components than other, even tropical, ecosystems. I mentioned several of these interactive links in the first section of this chapter, but more will be illuminated in the coming chapters. The Western Amazon is also intimately involved with the basic biogeochemical cycles that make the Earth a place full of life and thus tied to our shared human future.

The book contains both flooded (e.g. *várzea*, *igapó*) and unflooded (e.g. *terra firme*, white-sand, palm) Western Amazonian forests. Authors will prepare chapters that consist of reviews of what is known about their topic, of the research they have done, and of what research needs to be done in the future, often with a new conceptual and/or mathematical model. My organizing theme will be the structure, function and dynamics of the various kinds of forests found in the Western Amazon: how they were in the past, how they are changing today, and how they are likely to change in the future. Authors will ask if the drivers for those changes (e.g. climate change, human disturbances, tree recruitment mechanisms, stress) differ now compared to the past. Conceptual models will make use of tree replacement, or barriers to replacement, by individual trees within these forests (Myster 2012d). My concluding chapter will first summarize the results from each of the preceding chapters, then synthesize those results adding to the general conceptual model, and finally suggest future avenues for research.

Acknowledgements

I thank Naomi Schemm for her help in searching for research materials and the Fulbright Foundation for three research and teaching awards at universities in Ecuador and Peru.

References

Anderson, J.T., Rojas, T.S. and Flecker, A.S. (2009) High-quality seed dispersal by fruit-eating fishes in Amazonian floodplain habitats. *Oecologia*, **161**, 279–290.

Baker, T.R., Honorio-Coronado, E.N., Phillips, O.L. *et al.* (2007) Low stocks of coarse woody debris in a southwest Amazonian forest. *Oecologia*, **152**, 495–504.

Balslev, H., Luteyn, J., Ollgaard, B. and Holm-Nielsen, L.B. (1987) Composition and structure of adjacent unflooded and floodplain forest in Amazonian Ecuador. *Opera Botanica*, **92**, 37–57.

Bierregaard, R.O., Gascon, C., Lovejoy, T.E. and Mesquita, C.G. (2001) *Lessons for Amazonia: the ecology and conservation of a fragmented forest*, Yale University Press, New Haven, CT.

Brokaw, N.V.L. (1982) The definition of treefall gap and its effect on measures of forest dynamics. *Biotropica*, **11**, 158–160.

Brokaw, N.V.L., Zimmerman, J.K., Willig, M.R. *et al.* (2012) Chapter 5: Response to disturbance, in *A Caribbean Forest Tapestry: the multidimentional nature of disturbance and response* (eds N. Brokaw, T.A. Crowl, A.E. Lugo *et al.*), Oxford University Press, Oxford, UK, pp. 201–271.

Calle-Rendon, B.R., Moreno, F. and Cardenas-Lopez, D. (2011) Soil and forest structure in the Colombian Amazon. *Revista de Biologia Tropical*, **59**, 1307–1322.

Carson, W.P. and Schnitzer, S.A. (2008) *Tropical Forest Community Ecology*, Wiley-Blackwell, New York.

Chacon, N., Dezzco, N., Rangel, M. and Flons, S. (2008) Seasonal changes in soil phosphorous dynamics and root mass along a flooded tropical forest gradient in the lower Orinoco river, Venezuela. *Biogeochemistry*, **87**, 157–168.

Choo, J.P.S., Martinez, R.V. and Stiles, E.W. (2007) Diversity and abundance of plants with flowers and fruits from October 2001 to September 2002 in Paucarillo Reserve, Northeastern Amazon. *Peru Revisita Peru Biology*, **14**, 25–31.

Connell, J.H. (1971) On the role of natural enemies in preventing competitive exclusion in some marine animals and in rain forest trees, in *Dynamics of Populations* (eds P.J. denBoer and G.R. Gradwell), Center for Agricultural Publishing and Documentation, Wageningen, The Netherlands, pp. 298–310.

Dalling, J.W. and John, R. (2008) Seed limitation and the coexistence of pioneer tree species, in *Tropical Forest Community Ecology* (eds W.P. Carson and S.A. Schnitzer), Wiley-Blackwell, New York, pp. 242–253.

Daly, D.G. and Prance, G.T. (1989) Brazillian Amazon, in *Floristic Inventory of Tropical Countries* (eds D.G. Campbell and H.D. Hammond), Botanical Garden, Bronx, New York, pp. 401–426.

Davidson, E.A., de Aranjo, A.C., Artaxo, P. *et al.* (2010) The Amazon basin in transition. *Nature*, **481**, 321–328.

Duivenvoorden, J.F. (1995) Tree species composition and rain forest-environment relationships in the middle Caquetá area, Colombia, NW Amazonia. *Vegetatio*, **120**, 91–113.

Duivenvoorden, J.F., Balslev, H., Caveilier, J. *et al.* (2001) *Evaluacion de recursos vegetales no maderables en la Amazonia noroccidental*, Institute for Biodiversity and Ecosystem dynamics, Universiteit van Amsterdam, The Netherlands.

Dumont, J.F., Lamotte, S. and Kahn, F. (1990) Wetland and upland forest ecosystems in Peruvian Amazonia: plant species diversity in the light of some geological and botanical evidence. *Forest Ecology and Management*, **33/34**, 125–139.

Everham, E.M. III, Myster, R.W. and VanderGenachte, E. (1996) Effects of light, moisture, temperature and litter on the regeneration of five tree species in the tropical montane wet forest of Puerto Rico. *American Journal of Botany*, **83**, 1063–1068.

Falster, D.S., Duursma, R.A., Ishihara, M.I. *et al.* (2015) BAAD: a Biomass and Allometry Database for woody plants. *Ecology*, **96**, 1445.

Ferreira, L.V. and Stohlgren, T.J. (1999) Effects of river level fluctuation on plant species richness, diversity, and distribution in a floodplain forest in Central Amazonia. *Oecologia*, **120**, 582–587.

Fine, P.V.A., Mesones, I. and Coley, P.D. (2004) Herbivores promote habitat specialization by trees in Amazonian forest. *Science*, **305**, 663–665.

Frederickson, M.E., Greene, M.J. and Gordon, D.M. (2005) "Devil's garden" bedeviled by ants. *Nature*, **437**, 495–496.

Gentry, A.H. (1993) *A field guide to the families and genera of woody plants of northwest South America (Colombia, Ecuador, Peru) with supplementary notes on herbaceous taxa*, Conservation International, Washington, DC.

Gorchov, D.L., Palmeirim, J.M., Jaramillo, M. and Ascorra, C.F. (2004) Dispersal of seeds of *Hymenaea courbaril* (Fabaceae) in a logged rain forest in the Peruvian Amazonian. *Acta Amazonica*, **34**, 251–259.

Gottdenker, N. and Bodmer, R.E. (1998) Reproduction and productivity of white-lipped andcollared peccaries in the Peruvian Amazon. *Journal of Zoology, London*, **245**, 423–430.

Grubb, P.J. (1977) The maintenance of species richness in plant communities: the importance of the regeneration niche. *Biological Reviews*, **52**, 107–145.

Hecht, S.B. (2014) *The Scramble for the Amazon and the Lost Paradise of Euclides Da Cunha*, University of Chicago Press, Chicago, IL, p. 612.

Holdridge, L.R. (1967) *Life Zone Ecology*, Tropical Science Center, San Jose, Costa Rica.

Honorio, E.N. (2006) *Floristic relationships of the tree flora of Jenaro Herrera, an unusual area of the Peruvian Amazon*, MSc thesis, University of Edinburgh, Edinburgh, UK.

Hoorn, C. and Wesselingh, F.P. (2010) *Amazonia: Landscape and Species Evolution: A look into the past*, Wiley-Blackwell, Oxford, UK.

Janzen, D.H. (1970) Herbivores and the number of tree species in tropical forests. *Am. Nat.*, **104**, 501–528.

Janzen, D.H. (1984) *Costa Rican Natural History*. University of Chicago Press, Chicago, IL.

Jordan, C.F. (1985) Soils of the Amazon Rainforest, in *Amazonia* (eds G.T. Prance and T.E. Lovejoy), Pergamon Press, Ltd, Oxford, UK, pp. 83–94.

Junk, W.J. (1984) Ecology of the Varzea, floodplains of Amazonian white-water rivers, in *The Amazon: limnology and landscape ecology of a mighty tropical river and its basin* (ed. W.T. Junk), Kluwer, Dordrecht, pp. 215–243.

Junk, W.J. (1989) Flood tolerance and tree distribution in central Amazonian floodplains, in *Tropical Forests: Botanical dynamics, speciation and diversity* (eds L.B. Holm-Nielsen, I.C. Nielsen and H. Balslev), Academic Press, New York, pp. 47–64.

Junk, W.J. and Furch, K. (1985) The physical and chemical properties of Amazonian waters and their relationship with the Biota, in *Soils of the Amazon Rainforest. Amazonia* (eds G.T. Prance and T.E. Lovejoy), Pergamon Press, Ltd, Oxford, UK, pp. 3–17.

Junk, W.J., Piedade, M.T.F., Parolin, P. *et al.* (2010) *Amazonian Floodplain Forests: ecophysiology, biodiversity and sustainable management*, Ecological Studies, Springer-Verlag, Berlin.

Kalliola, R.S., Jukka, M., Puhakka, M. and Rajasilta, M. (1991) New site formation and colonizing vegetation in primary succession on the western Amazon floodplains. *Journal of Ecology*, **79**, 877–901.

Keller, M., Alencar, A., Asner, G.P. *et al.* (2004) Ecological research in the large-scale biosphere atmosphere experiment in Amazonia: early results. *Ecological Applications*, **14**, S3–S16.

Kern, J., Kreibich, M., Koschorreck, M. and Darwich, A. (2010) Nitrogen balance of a floodplain forest of the Amazon river: the role of nitrogen fixation, in *Amazonian Floodplain Forests: cophysiology, biodiversity and sustainable management* (eds W.J. Junk, M.T.F. Piedade, P. Parolin *et al.*), Ecological Studies, Springer-Verlag, Berlin, pp. 281–300.

Klinge, H., Junk, W.J. and Revilla, C.J. (1990) Status and distribution of forested wetlands in tropical South America. *Forest Ecology and Management*, **33/34**, 81–101.

Kubitzki, K. and Ziburski, A. (1994) Seed dispersal in flood plain forests of Amazonia. *Biotropica*, **26**, 30–43.

Lamotte, S. (1990) Fluvial dynamics and succession in the Lower Ucayali River basin, Peruvian Amazonia. *Forest Ecology and Management*, **33**, 141–156.

Langerwisch, R.F., Rosti, S., Gerten, D. *et al.* (2012) Potential effects of climate change on inundation patterns in the Amazon Basin. *Hydrology and Earth System Sciences Discussions*, **9**, 261–300.

Lopez, O.R. and Kursar, T.A. (1999) Flood tolerance of four tropical tree species. *Tree Physiology*, **19**, 925–932.

Losos, E.C. and Leigh, E.C. (2004) *Forest Diversity and Dynamism: Findings from a Network of Large-Scale Tropical Forests Plots*, University of Chicago Press, Chicago, IL.

Mabberley, D.J. (1992) *Tropical Rain Forest Ecology*, Chapman & Hall, New York.

Malhi, Y., Roberts, J.T., Betts, R.A. *et al.* (2008) Climate change, deforestation and the fate of the Amazon. *Science*, **319**, 169–172.

McClain, M.E., Victoria, R.L. and Richey, J.E. (2001) *The Biogeochemistry of the Amazon Basin*, Oxford University Press, Oxford, UK.

McDade, L.A., Bawa, K.S., Hespenheide, H.A. and Hartshorn, G.S. (1994) *La Selva: Ecology and Natural History of a Neotropical Rain Forest*, University of Chicago Press, Chicago, IL.

Melack, J.M. and Hess, L.L. (2010) Remote sensing of the distribution and extent of wetlands in the Amazon basin, in *Amazonian Floodplain Forests: ecophysiology, biodiversity and sustainable management* (eds W.J. Junk, M.T.F. Piedade, P. Parolin *et al.*)(eds), Ecological Studies, Springer-Verlag, Berlin, pp. 43–60.

Metz, M.R., Comita, L.S., Chen, Y.-Y. *et al.* (2008) Temporal and spatial variability in seedling dynamics: a cross-site comparison in four lowland tropical forests. *Journal of Tropical Ecology*, **24**, 9–18.

Meyer, U., Junk, W.J. and Linck, C. (2010) Fine root systems and mycorrhizal associations in two central Amazonian inundation forests: Igapo and Varzea, in *Amazonian Floodplain Forests: ecophysiology, biodiversity and sustainable management* (eds W.J. Junk, M.T.F. Piedade, P. Parolin *et al.*)(eds), Ecological Studies, Springer-Verlag, Berlin pp., pp. 163–178.

Montufar, R. and Pintaud, J. (2006) Variation in species composition, abundance and microhabitat preferences among western Amazonian *terra firme* palm communities. *Journal of the Linnean Society*, **151**, 127–140.

Moran, E.F., Brondizio, E.S., Tudor, J.M. *et al.* (2000) Effects of soil fertility and land-use on forest succession in Amazonia. *Forest Ecology and Management*, **139**, 93–108.

Moreau, C.S. (2008) Unraveling the evolutionary history of the hyperdiverse ant genus *Pheidole* (Hymenoptera: Formicidae). *Molenular Physiology Evolution*, **48**, 224–239.

Myers, N., Mittermeier, R.A., Mittermeier, G.C. *et al.* (2000) Biodiversity hotspots for conservation priorities. *Nature*, **403**, 853–858.

Myster, R.W. (2001) Mechanisms of plant response to gradients and after disturbances. *Botanical Review*, **67**, 441–452.

Myster, R.W. (2007a) *Post-agricultural Succession in the Neotropics*, Springer-Verlag, Berlin.

Myster, R.W. (2007b) Interactive effects of flooding and forest gap formation on composition and abundance in the Peruvian Amazon. *Folia Geobotanica*, **42**, 1–9.

Myster, R.W. (2009) Plant communities of Western Amazonia. *Botanical Review*, **75**, 271–291.

Myster, R.W. (2010) Flooding duration and treefall interactive effects on plant community richness, structure and alpha diversity in the Peruvian Amazon. *Ecotropica*, **16**, 43–49.

Myster, R.W. (2012a) *Ecotones between Forest and Grassland*, Springer-Verlag, Berlin.

Myster, R.W. (2012b) A refined methodology for defining plant communities using data after Sugarcane, Banana and pasture cultivation in the Neotropics. *The Scientific World Journal*, **2012** (365409) Article ID, 9.

Myster, R.W. (2012c) Spatial and temporal heterogeneity of light and soil water along a *terra firma* transect in the Ecuadorian Amazon. *Canadian Journal of Forest Research*, **42**, 1–4.

Myster, R.W. (2012d) Plants replacing plants: the future of community modeling and research. *Botanical Review*, **78**, 2–9.

Myster, R.W. (2013) The effects of flooding on forest floristics and physical structure in the Amazon: results from two permanent plots. *Forest Research*, **2**, 112.

Myster, R.W. (2014) Interactive effects of flooding and treefall gap formation on *terre firme* forest and várzea forest seed and seedling mechanisms and tolerances in the Ecuadorean Amazon. *Community Ecology*, **15**, 212–221.

Myster, R.W. (2015a) Várzea forest vs. *terra firme* forest floristics and physical structure in the Ecuadorean Amazon. *Ecotropica*, **20**, 35–44.

Myster, R.W. (2015b) Flooding × tree fall gap interactive effects on black-water forest floristics and physical structure in the Peruvian Amazon. *Journal of Plant Interactions*, **10**, 126–131.

Myster, R.W. (2015c) Comparing and contrasting eight different flooded and non-flooded forests in the Peruivan Amazon: seed rain. *New Zealand Journal of Forest Science*, **45**, 5.

Myster, R.W. (2016) Black-water forests (*igapo*) vs. white-water forests (*varzea*) in the Amazon: Floristics and physical structure. *The Biologist (Lima)*, **13**, 391–406.

Myster, R.W. (in press) The physical structure of Amazon forests: a review. *The Botanical Review.*

Myster, R.W. and Fetcher, N. (2005) Ecotypic differentiation and plant growth in the Luquillo Mountains of Puerto Rico. *Journal of Tropical Forest Science*, **17**, 163–169.

Myster, R.W. and Santacruz, P.G. (2005) Una comparación de campo de insectos de suelo-morar de Amazonas: Tierra firme y bosques de tierras inundadas vs. Espacios abiertos en el Parque Nacional Yasuní Ecuador. *Revista de la Pontificia Universidad Católica der Ecuador*, **76**, 111–124.

Myster, R.W., Lebron, L., Portugal-Loayza, A.B. and Zimmerman, J.K. (2013) Mycotrophic strategy of 13 common Neotropical trees and shrubs. *Journal of Tropical Forest Science*, **25**, 34–41.

Neff, T., Lucas, R.M., Dos-Santos, J.R. *et al.* (2006) Area and age of secondary forests in Brazilian Amazonia 1978–2002: An empirical estimate. *Ecosystems*, **9**, 609–623.

Nunez-Iturri, G. and Howe, H.F. (2007) Bushmeat and the fate of trees with seeds dispersed by large primates in a lowland rain forest in Western Amazonia. *Biotropica*, **39**, 348–354.

Paine, C.E.T. and Beck, H. (2007) Seed predation by Neotropical rain forest mammals increases diversity in seedling recruitment. *Ecology*, **88**, 3076–3087.

Paine, C.E.T., Harmes, K.E. and Ramos, J. (2009) Supplemental irrigation increases seedling performance and diversity in a tropical forest. *Journal of Tropical Ecology*, **25**, 171–180.

Parolin, P., DeSimone, O., Haase, K. *et al.* (2004) Central Amazonian floodplain forests: tree adaptations in a pulsing system. *Botanical Review*, **70**, 357–380.

Parolin, P., Wittmann, F. and Schongart, J. (2010) Tree phenology in Amazonian floodplain forests, in *Amazonian Floodplain Forests: ecophysiology, biodiversity and sustainable management* (eds W.J. Junk, M.T.F. Piedade, P. Parolin *et al.*)(eds), Ecological Studies, Springer-Verlag, Berlin pp., pp. 105–126.

Peixoto, J.M.A., Nelson, B.W. and Wittmann, F. (2009) Spatial and temporal dynamics of river channel migration and vegetation in central Amazonian white-water floodplains by remote sensing techniques. *Remote Sensing of Environment*, **113**, 2258–2266.

Phillips, O., L., Luiz, E.O.C. *et al.* (2009) Drought sensitivity of the Amazon Rainforest. *Science*, **323**, 1344–1347.

Pinedo-Vasquez, M., Ruffino, M.L., Sears, R.R. *et al.* (2011) *The Amazon Varzea: the decade past and the decade ahead*, Springer-Verlag, Berlin.

Pires, J.M. and Prance, G.T. (1985) The vegetation types of the Brazilian Amazon, in *Amazonia* (eds G.T. Prance and T.E. Lovejoy), Pergamon Press, Oxford, UK, pp. 109–145.

Poorter, L. (1999) Growth responses of 15 rain-forest tree species to a light gradient: the relative importance of morphological and physiological traits. *Functional Ecology*, **13**, 396–410.

Prance, G.T. (1979) Notes on the vegetation of Amazonia III. The terminology of Amazonian forest types subject to inundation. *Brittonia*, **31**, 26–38.

Prance, G.T. and Lovejoy, T.E. (1985) *Amazonia*, Pergamon Press, Ltd, Oxford, UK.

Puhakka, M., Kalliola, R., Rajasilta, M. and Salo, J. (1992) River types, site evolution and successional vegeation patterns in Peruvian Amazonia. *Journal of Biogeography*, **19**, 651–665.

Rice, A.H., Pyle, F.H., Saleska, S.R. *et al.* (2004) Carbon balance and vegetation dynamics in an oldgrowth Amazonian forest. *Ecological Applications*, **14**, S55–S77.

Romoleroux, K., Foster, R., Valencia, R. *et al.* (1997) Especies lenosas (dap → 1 cm) encontradas en dos hectareas de un bosque de la Amazonia ecuatoriana, in *Estudios Sobre Diversidad y Ecologia de Plantas* (eds R. Valencia and H. Balslev), Pontificia Universidad Catolica del Ecuador, Quito, Ecuador, pp. 189–215.

Ruokolainen, K., Tuomisto, H., Macia, M.J. *et al.* (2007) Are floristic and edaphic patterns in Amazonian rain forests congruent for trees, pteridophytes and Melastomataceae? *Journal of Tropical Ecology*, **23**, 13–25.

Russo, S.E. (2005) Linking seed fate to natural dispersal patterns: factors affecting predation and scatter-hoarding of *Virola calophylla* seeds in Peru. *Journal of Tropical Ecology*, **21**, 243–253.

Salati, E. (1985) The climatology and hydrology of Amazonia, in *Amazonia* (eds G.T. Prance and T.E. Lovejoy), Pergamon Press, Ltd, Oxford, UK, pp. 18–48.

Saleska, S.R., Didan, K., Huete, A.R. and da Rocha, H.R. (2007) Amazon forests green-up during 2005 drought. *Science*, **318**, 612.

Schongart, J., Wittmann, F. and Worbes, M. (2010) Biomass and net primary production of central Amazonian floodplain forests, in *Amazonian Floodplain Forests: ecophysiology, biodiversity and sustainable management* (eds W.J. Junk, M.T.F. Piedade, P. Parolin *et al.*)(eds), Ecological Studies, Springer-Verlag, Berlin pp., pp. 347–288.

Shukla, J., Nobre, C. and Sellers, P. (1990) Amazon deforestation and climate change. *Science*, **247**, 1322–1325.

Sioli, H. (1984) The Amazon and its main affluents: Hydrography, morphology of the river courses, and river types. *Monographiae Biologicae*, **56**, 127–165.

Soares-Filho, B. S., Nepstad, D. C., Curran, L. M., Cerqueira, G. C., Garcia, R. A. et al. (2006) Modelling conservation in the Amazon basin. *Nature* **440/23**, 520–523.

Spichiger, R., Loizeau, P., Latour, C. and Barriera, G. (1996) Tree species richness of a South-Western Amazonian forest (Jenaro Herrera, Peru, 73°40′W/4°54′S). *Candollea*, **51**, 559–577.

Svenning, J. (1999) Microhabitat specialization in a species-rich palm community in Amazonian Ecuador. *Journal of Ecology*, **87**, 55–65.

Terborgh, J. and Andresen, E. (1998) The composition of Amazonian forests: patterns at local and regional scales. *Journal of Tropical Ecology*, **14**, 645–664.

ter Steege, H.T., Pitman, N.C.A., Phillips, O.L. *et al.* (2006) Continental-scale patterns of canopy tree composition and function across Amazonia. *Nature*, **443**, 444–447.

Townsend-Small, A., McClain, M.E. and Brandes, J.A. (2005) Contributions of carbon and nitrogen from the Andes mounrtains to the Amazon river: evidence from an elevational gradient of soils, plants and river material. *Limnol. Oceanogr.*, **50**, 672–685.

Tuomisto, H., Ruokolainen, K. and Yli-Halla, M. (2003) Dispersal, environment, and floristic variation of Western Amazonian forests. *Science*, **299**, 241–244.

Valencia, R., Condit, R., Romoleroux, K. *et al.* (2004a) Tree species diversity and distribution in a forest plot at Yasuní National park, Amazonian Ecuador, in *Forest Diversity and Dynamism: findings from a network of large-scale tropical forests plots* (eds E.C. Losos and E.G. Leigh), University of Chicago Press, Chicago, IL, pp. 107–118.

Valencia, R., Condit, R., Foster, R.B. *et al.* (2004b) Yasuní forest dynamics plot, Ecuador, in *Forest Diversity and Dynamism: findings from a network of large-scale tropical forests plots* (eds E.C. Losos and E.G. Leigh), University of Chicago Press, Chicago, IL, pp. 609–620.

Valencia, R., Foster, R.B., Villa, G. *et al.* (2004c) Tree species distributions and local habitat variation in the Amazon: large forest plot in eastern Ecuador. *Journal of Ecology*, **92**, 214–229.

Valencia, R., Condit, R., Muller-Lamdau, H.C. *et al.* (2009) Dissecting biomass dynamics in a large Amazonian forest plot. *Journal of Tropical Ecolog*, **25**, 473–482.

Walter, H. (1973) *Vegetation of the Earth and the Ecological Systems of the Geo-Biosphere*, Springer-Verlag, New York.

Whitmore, T.C. (1989) *An Introduction Tropical Rain Forests*, Oxford University Press, Oxford, UK.

Wittmann, F.W., Junk, J. and Piedade, M.T.F. (2004) The *varzea* forests in Amazonia: flooding and the highly dynamic geomorphology interact with natural forest succession. *Forest Ecology and Management*, **196**, 199–212.

Wittmann, F., Junk, W.J. and Schongart, J. (2010a) Phytogeography, species diversity, community structure and dynamics of central Amazonian floodplain forest, in *Amazonian Floodplain Forests: ecophysiology, biodiversity and sustainable management* (eds W.J. Junk, M.T.F. Piedade, P. Parolin *et al.*)(eds), Ecological Studies, Springer-Verlag, Berlin.

Wittmann, A., Lopes, A., Conserva, A. *et al.* (2010b) Seed germination and seedling establishment on Amazonian floodplain trees, in *Amazonian Floodplain Forests: ecophysiology, biodiversity and sustainable management* (eds W.J. Junk, M.T.F. Piedade, P. Parolin *et al.*)(eds), Ecological Studies, Springer-Verlag, Berlin, pp. 259–280.

Wittmann, F., Householder, F.E., Piedade, M.T.F. *et al.* (2013) Habitat specifity, endemism and the neotropical distribution of Amazonian white-water floodplain trees. *Ecography*, **36**, 690–707.

Worbes, M., Klinge, H., Revilla, J.D. and Martius, C. (1992) On the dynamics, floristic subdivision and geographical distribution of *varzea* forests in Central Amazonia. *J. Veg. Sci.*, **3**, 553–564.

Zimmerman, J.K., Thompson, J. and Brokaw, N. (2008) Large tropical forest dynamics plots: testing explanations for the maintenance of species diversity, in *Tropical Forest Community Ecology* (eds W.P. Carson and S.A. Schnitzer), Wiley-Blackwell, New York, pp. 98–118.

2

A Floristic Assessment of Ecuador's Amazon Tree Flora

Juan E. Guevara, Hugo Mogollón, Nigel C. A. Pitman, Carlos Cerón, Walter A. Palacios and David A. Neill

Abstract

Ecuadorian Amazonia has been catalogued as one of the most biodiverse regions on Earth and is particularly renowned for the highest peaks of plant diversity. This chapter addresses the observed tree species diversity and the expected number of tree species in Ecuadorian Amazonia. It examines whether the Ecuadorian Amazon is floristically heterogeneous or forms a big block of relatively homogeneous forests. The chapter provides a brief overview of the landscape characteristics of the historical centers of botanical collection and research in Ecuadorian Amazonia. It considers these areas because of historic sampling efforts and the high intensity of botanic collection rather than any specific ecological analysis. The chapter discusses the role of geomorphology and soils on the patterns of floristic change in Ecuadorian Amazonia. Plant-herbivore interactions have been hypothesized as the main drivers of speciation and co-existence in Amazonian forests.

Keywords *ecological analysis; Ecuadorian Amazonia; floristic changes; homogeneous forests; landscape characteristics; plant diversity; plant–herbivore interactions; tree species diversity*

2.1 Introduction

Ecuadorian Amazonia has been catalogued as one of the most biodiverse regions on Earth (Bass *et al.* 2010; Funk *et al.* 2012; Pitman *et al.* 2001) and is particularly renowned for the highest peaks of plant diversity (de Oliveira and ter Steege 2013; Kraft *et al.* 2008; Kreft *et al.* 2004; ter Steege *et al.* 2013; Valencia *et al.* 2004). Located in the so-called Piedemonte del Napo region, this area is also characterized by the highest levels of tree and shrub diversity across the Amazon Basin (Fine and Kembell 2011; Myers *et al.* 2000; Orme *et al.* 2005; Pitman *et al.* 2001; ter Steege *et al.* 2013; Valencia *et al.* 1994, 2004).

Despite a long history of data collection, a complete assessment of the tree flora across the region has remained elusive for both ecologists and botanists during the past 50 years. Historically, the centers of botanical exploration have been

Forest structure, function and dynamics in Western Amazonia, First Edition.
Edited by Randall W. Myster.

Yasuní National Park (YNP) and to a lesser extent the Cuyabeno Reserve, where more than 50% of the collections have been carried out (Cerón and Reyes 2003; Jorgensen and León-Yánez 1999; Valencia *et al.* 1994, 2004). This has left certain regions located to the south and northeast of these areas fairly unexplored and known from just a few collections and small-scale censuses. This lack of information is responsible for the partial picture of tree gamma diversity in Ecuadorian Amazon forests.

The first attempt to determine both floristic and abundance patterns in Ecuador's Amazonian tree flora was the work developed by Pitman *et al.* 2001 (see also Macía and Svenning 2005). Some of the most interesting conclusions of this work are that:

1. Forests are dominated by a small subset of species that represents almost 70–80% of the total number of individuals (see also ter Steege *et al.* 2013); and
2. This small subset of species forms predictable oligarchies, *sensu* Pitman *et al.* (2001), over large tracts of forests, if environmental conditions (soils particularly) are relatively homogeneous across the landscape.

On the basis of these conclusions, they suggested that Ecuadorean forests might represent a large block of forest dominated by the same oligarchic species due to the relative homogeneous soil composition across the region (Pitman *et al.* 2001, 2008)

Here we present the results of an extensive 1-ha plot network and herbarium data compilation to assess for the first time the number of tree species and the floristic patterns of lowland Ecuadorian Amazonia's hyper-diverse forests. In this section we address three basic questions:

1. what is the observed tree species diversity in Ecuador Amazonia?
2. what is the expected number of tree species? and finally
3. is Ecuadorian Amazonia floristically heterogeneous or does it form a big block of relatively homogeneous forests?

Ultimately, we discuss some potential factors and mechanisms that might be influencing the patterns we found.

2.2 Methods

To delimit lowland Amazonia, we followed a hierarchical-nested approach used in the *Vegetation Map of Ecuador* (Sistema de Clasificación de Ecosistemas del Ecuador Continental: Ministerio del Ambiente del Ecuador 2013). This approach uses seven levels of classification to define ecosystems; for the purposes of our work, the first five levels are relevant because they include parameters such as geomorphology, temperature, precipitation, biogeography and topography at local and landscape scales to define the major biogeographic regions (see Sistema de Clasificación de Ecosistemas del Ecuador Continental, 2013).

2.3 Study area

In this section, we provide a brief overview of the landscape characteristics of the historical centers of botanical collection and research in Ecuadorian Amazonia. We have chosen these areas because of the historic sampling efforts and the high intensity of botanic collection rather than any specific ecological analysis. Unfortunately, this bias has been considered in previous attempts to define these regions *a priori* as floristically distinct.

2.3.1 Yasuní

The focus of botanical exploration has been the northwestern portion of the YNP (Bass *et al.* 2010; Pitman 2000; Pitman *et al.* 2001; Valencia *et al.* 2004). The great majority of the information collected during the past 15 years makes this portion of YNP well known in floristic terms, regardless of the fact that there are many other areas inside the park that remain unexplored. The YNP covers almost 980,000 ha with an altitude range from 175 to 400 m a.s.l. (Pitman 2000; Valencia *et al.* 2004). The landscape is dominated by hilly terrain, with 50-m high hills interrupted by small valleys drained generally by small streams that occasionally flood the adjacent areas. This strong and abrupt topographic variation is almost ubiquitous in the northwestern portion of YNP. Nonetheless, flat areas with poor drainage are predominant in the floodplains of the major rivers that cross the park and swampy areas are also scattered across the region (Guevara obs. pers.; Pitman 2000; Pitman *et al.* 2014; Valencia *et al.* 2004).

Terra firme and swamp forests are the predominant habitat types in YNP and cover almost 80% and 20% of the area respectively (Pitman *et al.* 2014; Sistema de Clasificación de Ecosistemas del Ecuador Continental 2013). Swamp forests in Yasuní are divided in two well-differentiated types that previously have not been recognized: habitats dominated by the palm species *Mauritia flexuosa* (e.g. "moretales") and the mixed swamp forests with higher species richness located in valleys adjacent to *terra firme* forests (Pitman *et al.* 2014).

2.3.2 Cuyabeno

The area between the Aguarico and Napo rivers, as well as the Cuyabeno and Lagartococha lacustrine systems, are part of Cuyabeno Reserve. This region is characterized by poorer soil conditions with respect to the Yasuní region or other areas of Ecuadorian Amazonia. On average, the clay:sand ratio in most of the landscape of the Cuyabeno Reserve is 2:8, while on average it is x:x in the Yasuní region is 48%. In addition, levels of aluminum are also higher in the Cuyabeno region compared with other areas of Ecuadorian Amazonia (Cerón and Reyes 2003; Guevara 2006; Poulsen *et al.* 2006; Saunders 2008).

The geomorphology in this region is remarkably different from the rest of the Amazonian lowlands in Ecuador. Most of the landscape is characterized by rolling plains with low hills that do not surpass 20 m in height. This area is also interrupted by small valleys, but the extent of swampy areas is low compared with the Yasuní region. Moreover, it has been reported that soil composition is similar between this area and areas farther east in Amazonia such as the Yavarí region in peru (Pitman *et al.* 2003). On the northern banks of the Aguarico River, the landscape is dominated by extensive non-inundated alluvial terraces, rolling plains and few interspersed swamp areas. The extensive terraces of recent origin (e.g. Pleistocene) close to the margin of the river and under flooding regimes are interrupted only by high terraces with flat surfaces that have not suffered erosion of their surfaces (Saunders 2008; Wesselingh *et al.* 2006).

2.4 Herbarium collections

A preliminary compilation research of species records was performed based on the *Catalogue of Vascular Plants of Ecuador* (Jorgensen and León-Yánez 1999). After this initial step, exhaustive research of collections of tree and shrub species collected in Ecuador's Amazonian lowland forests was carried out by Guevara from 2008 to 2013 from vouchers deposited in Herbario Nacional del Ecuador (QCNE), Catholic University Herbarium (QCA) and Alfredo Paredes Herbarium (QAP). We also include identified specimens cited in TROPICOS (http://www.tropicos.org) and material collected and cited in the virtual flora database of The Field Museum of Chicago (http://fm1.fieldmuseum.org/vrrc/). In both cases, we reviewed every specimen recorded in these databases to confirm that plant family circumscriptions follow APG III (APG III 2009; Chase and Reveal 2009). During this time we reviewed every voucher specimen of the previously recorded species and we added new records and corrected misidentifications. Most of the new records or new species have been confirmed by the specialists in each group, but in many other cases our extensive experience in Amazonian tree species identification allows us to be confident about the accuracy of the binomial name assignation. Together our experience in Amazonian tree species identifications totals more than 90 years.

2.5 Floristic inventories

In addition to the herbarium collections, we established a 70 1-ha plot network from 1980 to 2013 in Ecuadorian Amazonia lowland forests (Figure 2.1). This plot network includes the major habitat types across the region (e.g. *terra firme*, swamp mixed forests, palm-dominated swamp forests, *igapó* and *várzea*, white sands). Our plot network includes areas previously not visited by other botanical teams such as the lower portion of Cordillera del Condor and the Pastaza River watershed. In each plot we recorded, tagged and identified every single tree with diameter at breast height (dbh) above or equal to 10 cm. Botanical collections for every tree species were collected and duplicates were compared with and deposited as botanical specimens in the following herbaria: MO, QCNE, QAP, and F.

For every species recorded in the plot network and in the herbaria collections, we obtained information for their presence in Ecuadorian Amazonia, including information about records in every province of the Ecuador's Amazon region,

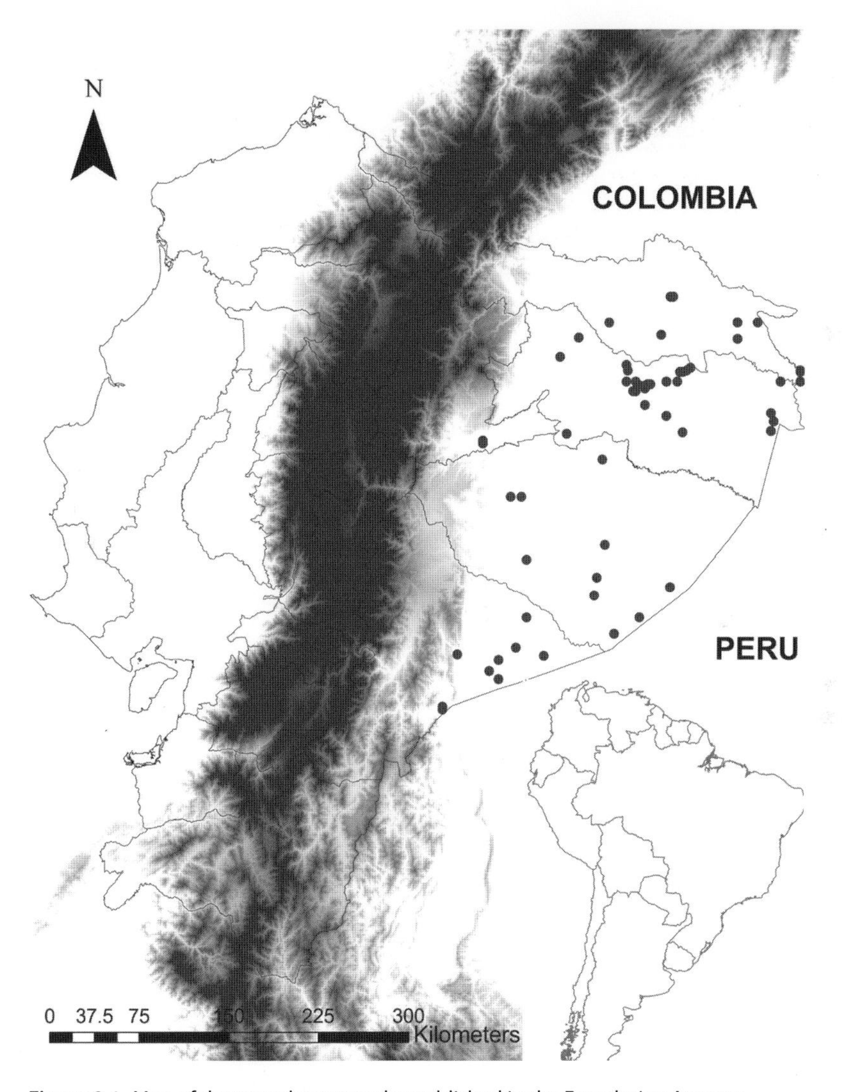

Figure 2.1 Map of the tree plot network established in the Ecuadorian Amazon.

habit, abundance and whether the species was previously recorded. All this information was gathered from the *Catalogue of Vascular Plants of Ecuador*.

2.6 Data analysis

We performed non-metric multi-dimensional analysis (NMDS) based on the Bray-Curtis dissimilarity index to establish the floristic relationships of the 1-ha plots we established across the Ecuadorian Amazon. NMDS is an ordination technique that forms groups of similar or dissimilar units in a bi-dimensional space based on abundance or presence-absence data. This technique allows us to group similar sampling units based on dissimilarity matrices without conserving the original distance, but giving the best solution for the ordination based on those matrices.

2.6.1 Estimation of observed and expected tree species richness

In order to determine the observed number of species in Ecuadorian Amazon, we followed a two-step process. First, we counted the number of tree species with valid names in our plot network. Second, we added these species to the tree species published in the *Catalogue of Vascular Plants of Ecuador* (Jorgensen and Leon-Yanez 1999), the *Addition to the Flora of Ecuador* (Neill and Ulloa 2011) and new records that have not been formally published.

To estimate the expected number of tree species, we followed a similar approached used by Pitman *et al.* (2002) and ter Steege *et al.* (2013). Fisher's alpha index (α) was preferred over other extrapolation methods, due to two basic assumptions: the first one implies that species abundance follows a log series distribution and the second that the regional species pool is spatially homogeneous. Based on the results published in ter Steege *et al.* (2013), the first assumption is fulfilled while the second assumption is still a matter of debate. However, based on previous evidence, this could be a good approximation for Ecuador's Amazonian forests (Pitman *et al.* 2001, 2008).

Fisher's alpha is calculated as:

$$s = \alpha \ln\left(1 + \frac{n}{\alpha}\right) \tag{2.1}$$

where S is the number of species and n is the number of individuals in any community or meta-community. As a consequence, knowing α and the population size, one can easily estimate the expected number of species.

In order to obtain an estimate the number of trees in the Ecuadorian Amazon tree population size, we extrapolated the number of individuals from the 70 1-h plot networks to the 6,901,090 ha of original forest (Sierra 2013). Then, assuming a forest loss of 11% in the last 25 years for the Amazon region below 500 m (Mapa Histórico de Deforestación: Ministerio del Ambiente del Ecuador 2013a), we obtained the number of trees in the 6,551,700 ha of remnant forest. With this information we calculated the expected number of tree species and compared it with the combined observed data calculated as the mean of the individual plots alphas. In addition, we used the Chao 2 estimator, which incorporates Hill numbers or the effective number of species approach (Chao *et al.* 2014). Because Fisher's α is an asymptotic estimator of species richness while Chao's is a diversity index, we considered it appropriate to compare these two approaches in order to obtain the most accurate estimate of the expected number of tree species in lowland Ecuadorian Amazon.

2.7 Results

2.7.1 Observed patterns of tree species richness

In herbaria collections and the tree plot network we recorded 2,183 species of trees with taxonomic valid names. Forty-four percent of all species (964 ssp) were already present in Jorgensen and León-Yánez (1999), while 56% (1,219 ssp) of all species with taxonomic valid names were present in the 1-ha tree plot network.

The three provinces with the highest numbers of records are Orellana, Sucumbíos and Pastaza, in that order. Orellana province with 1,848 tree species was the area with the highest number of records. Sucumbíos had 1,755 records, of which 110 species had been recorded only in this province, specifically in the Cuyabeno and Lagartococha lacustrine system and in the *terra firme* forests east of the Cuyabeno Reserve adjacent to Güeppí National Park in Peru.

Finally, the forests below 500 m in Pastaza province, with 708 records of tree species, is by far the least explored area in the Ecuadorian Amazon.

These results put in perspective the differences in sampling effort between areas adjacent to oil blocks or mining and those areas that remain inaccessible, even within a relatively well explored region such as Ecuadorian Amazonia.

The regional tree flora is characterized by a disproportionate number of taxa in families such as Fabaceae s.l. (222 ssp), Rubiaceae (163 ssp), Lauraceae (138 ssp), Annonaceae (110 ssp), Moraceae (110 ssp) and Euphorbiaceae (92 ssp) (Figure 2.2).

Low levels of endemism were detected in our analysis; approximately 3% of the total tree flora could be defined as endemic (Table 2.1). The families with the highest number of endemic species were Melastomataceae (15 ssp), Rubiaceae (10 ssp) and Fabaceae s.l. (7 ssp) (Figure 2.2). While some endemics are large trees, such as *Strypnodendron porcatum*, *Parkia balslevii*, *Andira macrocarpa*,

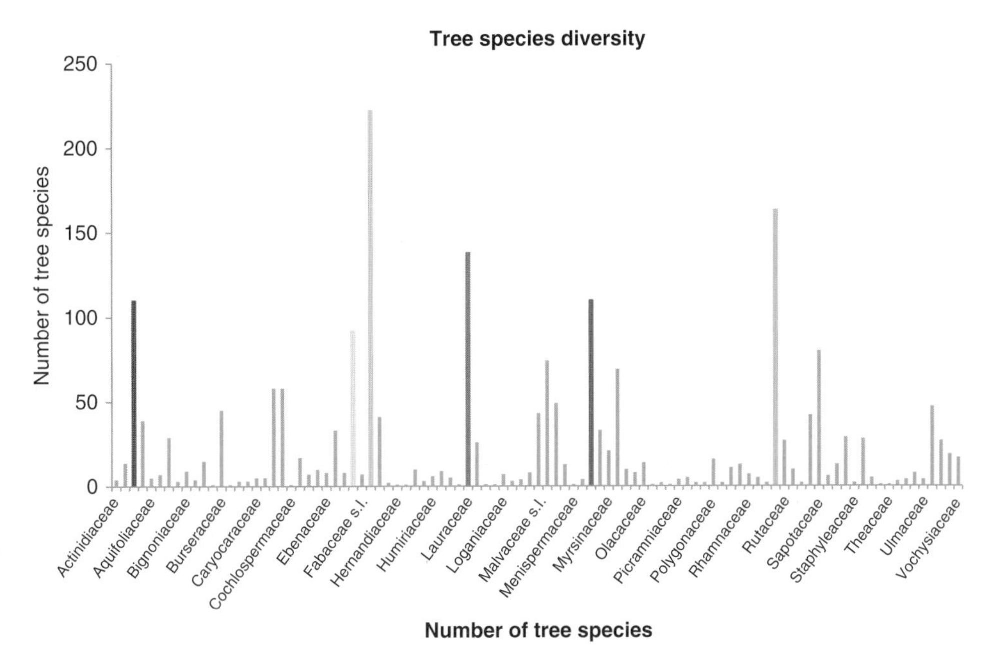

Figure 2.2 Patterns of diversity at the family level in the tree flora of the Ecuadorian Amazon.

Table 2.1 Estimates of tree species diversity, total number of trees and levels of endemism in Ecuador's Amazonian tree communities. Both values for Fisher's alpha and Chao's 2 metrics are shown.

	Original Area (ha)	Area remaining (ha)	Total number of trees	Fisher's alpha	Fisher's alpha(est)	Chao 2 Mean	Chao 2 95% CI Lower Bound	Chao 2 95% CI Upper Bound	Percent species that are endemic
Amazonian lowland forest	6.901.090	6.551.700	3.332.267.417	160.2	2.696	3.121	2.963.4	3.261.5	3.1

Swartzia bombycina, S. aurosericea and *Pentaplaris huaronica,* the majority of endemic species were small trees.

Both endemic trees and treelet species may exhibit small range size, restricted dispersal and sometimes clumped distributions at local scales, probably due to habitat specificity (see Duque *et al.* 2002; Guevara 2006; Pitman *et al.* 2001; Valencia *et al.* 2004). Particularly interesting is the fact that almost 90% of endemic species have been recorded close to the Andean foothills (Neill and Ulloa-Ulloa, 2011). In the cases, the endemics exhibit high relative abundance at local scales, in areas with unusual edaphic and geological particular conditions. For instance, the emergent endemic tree *Gyranthera amphibiolepis* (Malvaceae s.l.) has been reported in a narrow altitudinal band from 850 to 1,500 m on the eastern slopes of the Andes in Ecuador and Peru on calcareous limestones.

Five percent of the tree flora has also been recorded in Ecuador as treelets, while just 1% has been recorded as shrubs. In some cases these species change habit in response to changes in environmental gradients. A remarkable example of such plasticity is *Memora cladotricha* Sandwith, which can grow at the same locality as a tree or as a liana, depending on light availability.

2.7.2 Estimated number of tree species in Ecuadorian Amazonia

Based on our calculations, more than 3.3 billion adult trees co-exist in the Ecuadorian Amazon. We think our calculations are reasonable considering the 10 billion trees estimated for all of Ecuador (Crowther *et al.* 2015). Our estimate of tree species based on Fisher's α resulted in a mean of 2,696 species that may co-exist in the lowland Amazonian forests of Ecuador (Figure 2.3, Table 2.1). When we take into account the expected number of tree species based on Hill numbers, the estimate is significantly higher. Chao's estimation of tree species resulted in a mean of 3,121 tree species, almost 300 tree species more than the Fisher's α estimate.

The difference between estimates of tree species richness based on Fisher's α and Chao's algorithms may reflect the nature of both metrics. For instance, it is widely known that Fisher's α is a scale-independent non-parametric estimator that has a good discriminatory power to detect richness under the assumptions that communities sampled are characterized by a non-clustered spatial distribution of species and number of species tends to infinity (Schulte *et al.*

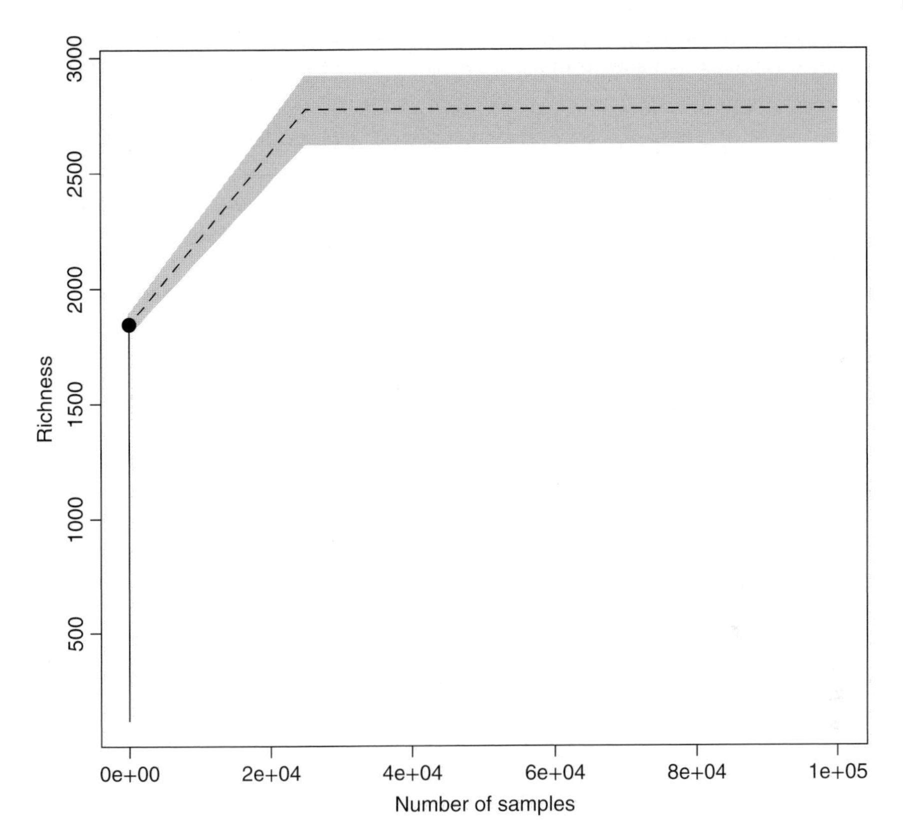

Figure 2.3 Species area curve and the estimated number of tree species expected for Ecuadorian Amazonia. Gray shaded area represents the 95% upper and lower bounds for the expected number of species with respect to area sampled. Data are based on a plot network of 70 one-hectare plots established across Ecuadorian Amazonia.

2005). In such cases, α might be underestimated. However, this could not be the case for the Ecuadorian Amazon, despite the fact that we assume *a priori* tree communities in this region follow this pattern. Meanwhile, Chao's 2 estimator works under the concept of the effective number of species (Hill numbers), which has been demonstrated to be a function of increasing sample effort and sample completeness.

Previous works have suggested an estimate of 3,370 species of woody plants including trees (Pitman *et al.* 2002). The most recent attempt to determine the expected number of tree species in the Ecuador Amazon predicts 6,827 tree species to occur in these hyperdiverse forests. This estimate was based on a fitting of tree species rank-abundance data to Fisher's log series distribution (ter Steege *et al.* 2013). However, ter Steege used a spatial model, with a relatively coarse span (1 degree grid cells = 111 km), to estimate the number of species based on the species rank-abundance curve. This could have resulted in the inclusion of areas that correspond to the Andean foothills above 500 m, which is considered the upper limit for lowland Amazonia. Therefore, we argue that this number might represent an overestimation of the real number of tree species in Ecuador Amazon. This assumption is supported by the fact that increasing sampling area the probablity of estimating a higher number of tree species will increase as well. In the case of ter Steege's estimation, this would result in the increase of the number of potential unsampled species in the right tale of the RAD distribution curve.

Our results could be interpreted as an underestimation of tree species diversity based on the work of Pitman *et al.* (2002). Nonetheless, we argue that this difference is artificial and lies in several factors. First of all, Pitman *et al.* (2002) estimates a regional species pool based on a sample of individuals with dbh $\geq$ 1 cm, increasing the probability of sampling not just juveniles of the species that we could not record in our tree inventories, but also species with other life forms. In the list of species published by Romoleroux *et al.* (1997), the cornerstone for Pitman's analysis, one can find species of free-standing lianas, juveniles of treelets and vines and some species that can exhibit more than one life form. In addition, estimates of regional diversity following Pitman's approach do not consider habitats other than *terra firme* forests and were built upon a small sample size (2 ha). Differences in α diversity in contrasting environments (e.g. *terra firme* vs. swamp forests or *terra firme* vs. white-sand forests) have been well documented and they are one of the reasons why incorporating low tree diversity habitats into the analysis should lead to lower-overall Fisher's α values and consequently a lower than predicted regional species pool. Furthermore, we must consider that since the last time an estimate for woody plant species richness was done in Ecuador Amazon, there has been a reduction of almost 450,000 hectares of native forest (Sierra 2013; Pitman *et al.* 2002). Deforestation has direct negative effects on the overall number of Amazon trees, but particularly on the population sizes of rare, small-ranged tree species. Because the expected number of tree species is a function of the total number of trees, we consider our estimates to be accurate considering the regional context of deforestation.

While to our knowledge this is the first attempt to estimate regional tree species diversity based on an intensive sampling across Ecuadorian Amazonia, the number of species we present is just a proxy of the actual tree diversity. Based on the results discussed above, we can predict that approximately 3400 tree species with dbh above 10 cm might occur in Ecuador's Amazon lowland forests.

2.7.3 Floristic relationships and discontinuities at local and regional scales

Based on herbaria collections and the results from the NMDS (stress function = 0.145) based on plot data, we are able to define four floristically distinct sub-regions in Ecuadorian Amazonia. These are: (1) the Cuyabeno-Güeppí region and the interfluvial area between the Napo and Aguarico rivers (Aguarico-Putumayo-Caqueta watershed), (2) the Napo-Curaray watershed, (3) the Pastaza watershed and (4) the lowland forests adjacent to Cordillera del Cóndor (Figure 2.3). We have named these biogeographic sub-regions based on the floristic differences between them, their floristic affinities with other Amazonian floras, and considering previous attempts that have incorporated geology, digital elevation models and climatic variables (see Sistema de Clasificación de los Ecosistemas del Ecuador Continental 2013).

2.8 Aguarico-Putumayo watershed

We found that the region of Güeppí-Cuyabeno has many elements of floras in Central Amazonia, including areas adjacent to the white-sand forests of the Iquitos and the Middle Caquetá region, the latter with a strong floristic influence of Guiana Shield forests. Genera such as *Sterigmapetalum, Chaunochiton, Neoptychocarpus, Macoubea, Podocalyx, Pogonophora, Antrocaryon, Botryarrhena, Clathrotropis, Adiscanthus, Ruizterania* and *Neocalyptrocalyx* have been recorded only in the interlying region between the interfluve of the Aguarico and Putumayo rivers and in the hill forest into the *terra firme* and on the high terraces of these two rivers' land.

Data from the network plots show us that the most diverse families in this region are, in order, Fabaceae s.l., Burseraceae, Chrysobalanaceae, Sapotaceae, Annonaceae, Lecythidaceae and Moraceae. In terms of abundance, the Burseraceae, Myristicaceae and Lecythidaceae are the most abundant families contrasting with the diversity patterns in forests on more fertile soils such as those ones located in Yasuní and near the foothills of the Andes, in which the diversity and abundance of families Moraceae, Arecaceae, Sapotaceae and Fabaceae s.l. are considerably higher. *Oenocarpus bataua, Neoptychocarpus killipi, Sterculia killipiana, Roucheria schomburgkii, Hebepetalum humirifolium, Swartzia racemosa, Dacryodes chimantensis, Couratari oligantha, Tachigali setifera, Eschweilera itayensis, E. rufifolia, Iryanthera lancifolia* and *I. laevis, Virola elongata, Licania octandra, L. cuyabenensis* and *Protium* have been reported with high local abundance in plots located in the Cuyabeno Güeppí region. Species of trees such as *Macoubea guianensis, M. sprucei, Rhigospira quadrangularis* and *Erythroxylum divaricatum* among others only have been recorded in our plots in this region. In several cases, the number of collections deposited in herbaria is much smaller than the abundance we have recorded in the plots. For example, *S. racemosa, C. oligantha, S. killipiana* and *N. killipi* are poorly represented in herbaria, but abundant in several of our plots located in the easternmost portion of the Ecuadorian Amazon.

Most of the new records in the Güeppí Cuyabeno area include species that are locally abundant in areas of the Middle Caquetá and certain areas near Manaus (De Oliveira and Daly 1999, De Oliveira and Mori 1999; Duivenvoorden 1995; Duque *et al.* 2002; Pitman *et al.* 2003). Thus, we think the northeastern portion of the Ecuadorian Amazon, the region of the triple border between Ecuador, Colombia and Peru, is the westernmost edge of Amazon with floristic influence of the Central Amazonia and Guiana Shield region. In the NMDS ordination, these plots established in the Aguarico-Putumayo–Caquetá region form a completely separate set of plots from the northwestern area of Yasuní.

2.9 Napo-Curaray basin

The Napo-Curaray basin region is characterized by high levels of tree alpha diversity, and the results of the tree inventory show that peaks of tree diversity can be found in the forests located in the YNP area (Figure 2.5). Some groups, such as Arecaceae, Fabaceae, Moraceae, Rubiaceae, Sapotaceae and Melastomataceae are remarkably dominant in terms of abundance in this region. Fabaceae, Lauraceae, Myrtaceae, Rubiaceae, Melastomataceae and Sapotaceae exhibit high levels of diversity as well. Moreover, *Inga, Ocotea, Pouteria, Virola, Eugenia* and *Calyptranthes* are species-rich genera that exhibit peaks of diversity in the YNP.

Towards the south of this region, in the Curaray and Tigre river basins, most of the forests remain unexplored. Very few botanical works have been done in this vast region and the probability of finding new species and dozens of new records is high. However, based on the limited information available in herbarium collections and the results of the NMDS ordination and the indicator species analysis (ISA) (Appendix, Table 1), this region shares floristic affinities with the flora of Loreto in Peru, particularly with forests located on sandy-clayey soils in the upper Nanay and the high terraces of the Napo river close to Iquitos. *Marlierea umbraticola, Arachnotryx peruviana, Pterygota amazonica, Huberodendron swietenioides, Cinchonopsis amazonica, Vatairea erythrocarpa* and *Ardisia huallagae* are conspicuous elements of the tree flora in this region. In addition, we recorded characteristic elements of the Central Brazilian Amazon. Species such as *H. humirifolium, Cassipourea guianensis, Licaria aurea, Septotheca tessmannii, Cleidon amazonicum, Compsoneura lapidiflora, Conceveiba martiana* and *Hura crepitans* grow between high and low terraces and upland forests of the Tigre river basin.

The abundance data of the three plots established near the border with Peru show that in certain areas the flora of Yasuní, on richer soils, overlaps with elements of the flora of the Alto Nanay and high Arabela areas, with a mixture of white-sand and sandy-clayey soil species.

2.10 Pastaza basin region

Until a few years ago, the Pastasa basin region had remained unexplored. Our results demonstrate that the forests to the north of the Morona river and south of the Pastaza river change in species composition compared with the forest located on the adjacent sandstones at higher and lower elevations in El Cóndor (Figures 2.4 and 2.5). The confluence of several floras, including widely-distributed elements of the Amazon piedmont, the flora of Guyana Shield forests and the region of Iquitos on mixed soils, determines the patterns we found. Some Guyana Shield elements include *Chlorocardium rodiei, Zapoteca amazonica, Cochlospermum orinocensis* and *Calycophyllum obovatum* and species in the region of Loreto such as *Vantanea peruviana, Acacia loretensis, Ruizterania trichantera, Chrysochlamys ulei* and *H. crepitans* have been also recorded in these plots and seem to be characteristic of the flora of this region.

In text, it is clear that the Ecuadorian Amazon is more heterogeneous than previously hypothesized. The overlapping of several regional floras in a relatively small area such as lowland Amazonian Ecuador proposes a new perspective on patterns of relative abundance and species diversity. The tree and shrub

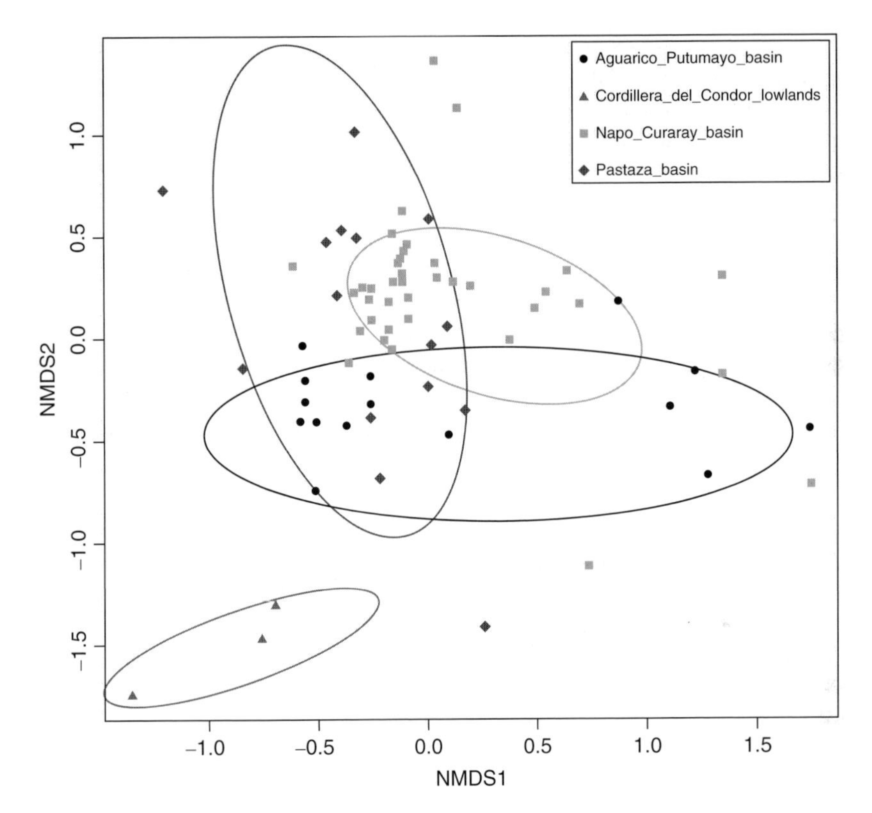

Figure 2.4 Non Metric Multidimensional ordination based on the species-level compositional dissimilarity matrix for 70 1-h plots in *terra firme* forests of the Ecuadorian Amazon. Ellipses represent the 95% confidence interval in grouping plots as part of a particular cluster of similar floristic units. Four floristically distinct regions are defined: triangle ellipse = forests in the lowlands of the Cordillera del Cóndor region; black ellipse = forests in the Aguarico-Putumayo watershed; square ellipse = forests in the hyperdiverse Napo-Curaray watershed; and diamond ellipse = forests in the Pastaza watershed.

community in the Ecuadorian Amazon reflects the influence of a number of forces, including historical evolutionary processes to current ecological processes, such as limited niche assembly and dispersal, the latter two believed to have happened simultaneously at different spatial scales (Antonelli *et al.* 2009; Cavender-Bares *et al.* 2009; Pennington and Dick 2010; Ricklefs 2006).

2.11 Cordillera del Cóndor lowlands

Finally, the forests below 500 m adjacent to the Cordillera del Cóndor and Cutucú could be considered as one of the areas of the Ecuadorean Amazon that remain floristically unexplored. A small number of works have focused on the floristic patterns of sandstone formations (called tepuis for their resemblance to Guyana Shield tepuis) and forests above 800 m; however, the forests below 500 m remain unexplored (Foster and Beltrán 1997; Jadan and Aguirre 2011; Neill and Ulloa 2002).

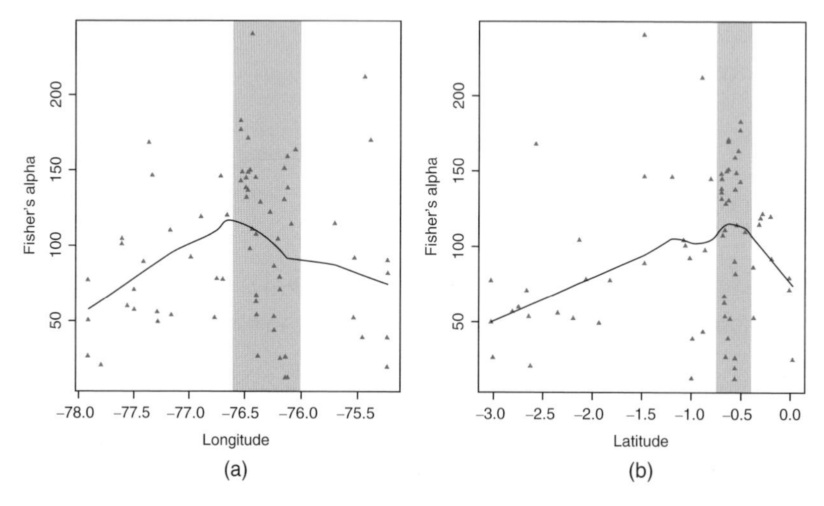

Figure 2.5 Tree alpha diversity (measured as Fisher's α) as a function of (a) longitude and (b) latitude. Loess regression is fitted to the data. The highest values of tree alpha diversity are found in the YNP region close to the Andean foothills and are probably related to climate and soil conditions. The gray bars highlight the Yasuni Park area.

In these studies, the authors have reported the presence of at least 16 genera with disjunct distributions that involve the Guyana Shield tepuis. Species such as *Stilnophyllum grandiflorum, Remija chelomophylla, Sterigmapetalum obovatum, Pagamea duddleyi* or *Dendrotrix yutajensis* and *Dacryodes uruts-kunchae* are diagnostic of forests on sandstone plateaus above 1,000 m in the Cóndor–Cutucú Cordillera. We believe, based on floristica data that comes from the few plots we established in the lowest part of this region, that some of the taxa mentioned above are part of the forests on sandstones below 500 m (Guevara *et al.* unpublished data). Moreover, some genera dominant in the white-sand forest of the surroundings of Iquitos and the upper Morona river watershed are also dominant in the low-elevation plateaus of El Cóndor Cordillera. Some of these genera include *Pachira, Micrandra, Diclinanona, Parkia* and *Ilex.*

Despite the strong floristic patterns we report, there is a substantial number of plots (>50%) that lie in more than one cluster, representing overlap between the proposed biogeographic regions (Figure 2.4). This result might contradict the patterns of strong floristic turnover we describe; however, this pattern is in agreement with previous results that posit a non-mutually exclusive presence of strong floristic dissimilarities and dominance in Amazon tree communities (Pitman *et al.* 2014). Due to Bray-Curtis tending to preserve the abundance of species at each site or local community to obtain distances between them, the more abundant species should have a strong influence on the results of an NMDS ordination if these are widespread across the landscape. In other words, even if rare species are not shared between pairs of local communities, strong floristic similarities could arise as the result of the presence of widespread abundant species.

2.12 What factors drive gradients in alpha and beta diversity in Ecuador's Amazon forests?

2.12.1 Climate and latitudinal and longitudinal gradients

We observed a unimodal longitudinal gradient in the tree alpha diversity of *terra firme* forests, while species richness along latitude shows a unimodal gradient (Figure 2.4a). Forests close to the Andean foothills are remarkably more diverse with respect their counterparts to the east of the basin. However, a latitudinal gradient also appears to be strong in terms of tree alpha diversity, as the results of the Generalized Least Squares (GLS) analysis demonstrated (Table 2.2). We found the highest AIC when considering the interaction of Longitude, Latitude and Climate, therefore this was not considered as the optimal model. Longitude alone is implausible as the only predictor for the changes we observed in tree alpha diversity; climate shift along this longitudinal gradient appears to be the best predictor for this trend instead, as observed in the GAM results (Figure 2.5; Table 2.3).

Latitudinal gradients in species richness have been extensively described in the literature (Fine 2015; Jablonski *et al.* 2006; Mittelbach *et al.* 2007; Rohde 1992). This general pattern is observable in several taxa including mammals, plants, insects and marine biota, and the trend appears to be stronger at continental and regional scales (Mittlebach *et al.* 2007). This astonishing trend, nonetheless, has not been documented at smaller spatial scales including sub-regional and landscape scales. One of the reasons for this lack of information lies in the fact that processes such as extinction, speciation and dispersal are fundamental in the determination of the regional species pool. Therefore being more predominant at larger spatial scales.

Table 2.2 Generalized Least Squares analysis (GLS) comparing monthly mean precipitation values, maximum, minimum and mean temperature values condensed in the axis 1 of a Principal Components analysis (PC1), with respect Fisher's alpha values, Longitude and Latitude. Model selection to obtain the best model possible was performed applying a backward selection and a likelihood ratio test. Significance of the variables interactions for each model is coded as: * = significant at the 5% level; ** = significant at the 1% level.

	Model	df	AIC	BIC	logLik	Test	L.Ratio	p-value	
Fisher_alpha~ Longitude+ Latitude+PC1	1	9	762.6473	782.1	−372				
Fisher_alpha~ PC1+Longitude	2	4	764.7193	773.7	−378	1 vs 2	12.07	**0.034**	*
Fisher_alpha~ PC1+Latitude	3	4	762.3509	771.3	−377				
Fisher_alpha~ PC1	4	3	769.1661	775.9	−382	3 vs 4	8.815	**0.003**	**

Table 2.3 Generalized Additive Model analysis (GAM) comparing climatic variables condensed in the axis 1 values of a Principal Components analysis (PC1), with Fisher's alpha values, Longitude and Latitude, to detect patterns of tree diversity in Ecuadorian Amazon forests. Significance of the variables interactions for each model is coded as: * = significant at the 5% level; ** = significant at the 1% level.

	Model	Estimate	Std. Error	t value	p-value	
(Intercept)		−3.52E+03	1.94E+03	−1.814	0.0743	*
PC1	1	1.75E+01	7.68E+00	2.276	0.0262	**
Longitude	2	4.78E+01	2.54E+01	1.88	0.0647	*
Latitude	3	2.78E+03	2.40E+03	1.158	0.2511	
PC1~Longitude	4	−2.28E−01	1.00E−01	−2.276	0.0262	**
PC1~Latitude	5	−1.03E+01	6.23E+00	−1.648	0.1042	
Longitude~Latitude	6	−3.67E+01	3.14E+01	−1.171	0.246	
PC1~Longitude+Latitude	7	1.34E−01	8.11E−02	1.655	0.1027	

We found that shifts in tree species diversity along latitudinal and longitudinal gradients can occur at smaller spatial scales, if environmental heterogeneity is strong enough to filter species from one region to another of the ecological space distributed over geographic space.

Examples of such longitudinal gradients occurring at smaller spatial scales than those repeatedly reported in the literature could occur in the Andean-Amazon transition from montane to lowland forests. In a study of aquatic communities across an Andean-Amazon fluvial gradient, Lujan *et al.* (2013) found a strong and non-linear shift in taxonomy and function of these communities and argued that strong turnover of basal food resources and inter-specific interactions was responsible for the change.

Although previous works have emphasized relatively homogeneous environmental conditions across lowland Amazonian in Ecuador (see Pitman *et al.* 2001; Tuomisto *et al.* 2003a), with respect to the more heterogeneous landscapes such as the forests surrounding Iquitos or the middle Caquetá, we believe that there is recent evidence to contradict this hypothesis.

Evidence for shifts in temperature and rainfall in an eastward direction has been reported for Ecuadorian Amazonia and could be responsible for the changes we detect, not just in diversity but also in composition (Figures 2.5 e,f).

Previously, Pitman *et al.* (2008) reported changes in tree species composition along a longitudinal gradient in a plot network from Yasuní to Yavarí along the Peruvian-Brazilian border. Strong and abrupt floristic changes were assumed to be produced by shifts in soil composition from richer soils in plots located in the Ecuadorian Amazon and poorer sandy soils in plots located on the northern bank of the Napo River.

We confirm these results but also posit that strong beta diversity along longitude might be correlated with climate and not just soil conditions.

2.13 The role of geomorphology and soils in patterns of floristic change in Ecuadorian Amazonia

There is an increasing number of studies showing the role of geology, geomorphology and soil heterogeneity in patterns of floristic change (beta diversity) at both landscape and regional scales in Amazonian forests (Figuereido *et al.* 2014; Fine *et al.* 2005; Higgins *et al.* 2011; Philips *et al.* 2003; Tuomisto and Poulsen 1996; Tuomisto *et al.* 2003a,b).

Higgins *et al.* (2011) reported a strong correlation between geological formations and their underlying soil conditions and fern and shrub species composition. Discontinuities in species composition therefore are tightly determined by geological history, which determines that forests are partitioned into large discrete floristic units. These changes instead follow a longitudinal gradient representing a complex history of deposition and erosion that predates the Pleistocene (Higgins *et al.* 2011; Hoorn *et al.* 1995; Hoorn 1996).

A strong correlation between abrupt floristic discontinuities, landforms and geological features has been also found in Central Amazonia. However, some caveats must be considered when interpreting this evidence. In both works, the authors test the role of geological control on Amazonian plant communities; however, the evidence relies on analysis done with groups of plants (Pteridophytes and Melastomataceae in Higgins *et al.* (2011) and Zingiberaceae in Figuereido *et al.* (2014)), which represent just a small fraction of total diversity. In both studies, the authors argue that focusing on these groups of plants helps to avoid taxonomic uncertainty, tall taxa and larger sample sizes, which involve a huge investment of time to demonstrate the patterns they reported. Although the authors predict the same patterns for the tree flora, we think this has to be tested instead of being assumed.

We found evidence for a correlation between geomorphology, soils and beta diversity in the Amazonian tree community. Towards the border with Peru, floodplains of the Napo, Putumayo and Aguarico rivers form high and low terraces with a fairly flat topography. The forest in this areas shows floristic affinities with plots established further east in the basin, registering families and monotypic genera present in areas of Central Amazonia (Alverson *et al.* 2008; Pitman *et al.* 2008).

For example, samples collected by Cerón and Montalvo (2009) in alluvial terraces of the Aguarico River correspond to species of monotypic genera such as *Podocalyx loranthoides* and *Pogonophora schomburgkiana*. These species are characteristic of poor soil forests in the Middle Caquetá and certain areas in Central Amazonia (Duivenvoorden 1996; Duque *et al.* 2002). In addition, in a rapid survey conducted in 2008 by the Field Museum in Chicago area in the Güeppí region, W. Palacios recorded the species *Chaunchiton kaplerii* (Alverson *et al.* 2008), a genus previously unknown for Ecuadorian Amazonia but with a widespread distribution.

Strong floristic affinities of forests along high terraces bordering the Napo River with forest eastwards of the Amazon Basin show a high abundance of species such as *R. quadrangularis, Mezilaurus sprucei, Rauvolfia polyphylla,*

Eriotheca longitubulosa, Sloanea monosperma and *V. peruviana* and high diversity and abundance of the genus *Protium*.

At the landscape scale, soils in this region are characterized by a very low pH and very high content of sand (Cerón and Reyes 2003; Saunders 2008).

Interestingly, studies in the Cuyabeno region have shown the predominance of certain taxa that are locally dominant and diverse in areas with poorer soils such as the sandy-clayey areas surrounding Iquitos (Fine *et al.* 2010; Valencia *et al.* 1994). Exceptional cases of family-level hyperdominance (*sensu* ter Steege *et al.* 2013) at the local scale (<1 km^2) have been reported in the Ecuadorian Amazon, such as the unusual dominance of Burseraceae in an upland forest in the Cuyabeno region (Cerón and Reyes 2003; see also Vormisto *et al.* 2004).

2.14 Potential evolutionary processes determining differences in tree alpha and beta diversity in Ecuadorian Amazonia

Despite the increasingly number of works analyzing and describing patterns of alpha and beta diversity in Amazonian forests, a comparatively small fraction of these works have included an evolutionary perspective in their analysis. We think that differences in tree species richness and tree species turnover are intimately linked to evolutionary processes that operate at large spatial scales. Differences in species richness across a particular region may be the result of asymmetries in speciation and extinction rates. These asymmetries could be the result of multiple factors, including area population size relationships, geographic barriers promoting disruption in gene flow between populations, colonization of novel habitats, habitat specialization, floral and pollinator specialization, herbivory and plant defense mechanisms, low dispersal capabilities, productivity or high energy systems (Fine 2015).

For example, diversification rates might differ between two areas if there is colonization of areas outside the ancestral range by high dispersal capability clades. High extinction rates could be the result of physiological constraints if new colonizers do not evolve a particular trait or set of traits that allows adaptations to the new habitat. Conversely, high speciation rates might be produced by high phenotypic plasticity that allows successful colonization of novel habitats, as lineage divergence in this sense is produced by local adaptations.

On the other hand, population size and geographic range size may promote speciation and extinction rate asymmetries. For instance, large population size might involve more genetic diversity if genetic drift is not involved, because speciation rates should increase while extinction rates decrease (Fine 2015; Jetz and Fine 2012). Moreover, species with large population sizes can span different climatic or other environmental regimes, which increases the probability of parapatric or peripatric speciation (Kisel *et al.* 2011).

Here we described some examples of works which have tested plausible mechanisms for asymmetries in speciation and extinction in Amazonian forests, which could be responsible for the longitudinal and latitudinal gradients.

Plant–herbivore interactions have been hypothesized as the main drivers of speciation and co-existence in Amazonian forests (Coley and Kursar 2014; Fine *et al.* 2004; Kursar *et al.* 2009). Selection for divergence in anti-herbivore defenses should lead to differences in speciation rates, if different herbivore guilds exert selective pressure differentially across a geographic region. In addition, separate populations in different environments should exhibit different adaptations that eventually would result in reproductive isolation when the members of these populations re-unite and interbreed. In this case, hybrids may have lower fitness than members of the ancestral populations accentuating selection for reproductive isolation (Coley and Kursar 2014).

Fine *et al.* (2005) demonstrated that soil heterogeneity leads to edaphic specialization and consequently drives speciation. In this study, the authors reported that 21 of 25 species of the tribe Protiae in Burseraceae were confined to one edaphic habitat. In particular, the great majority of these specialists apparently have evolved habitat specialization very recently. Moreover, the role of historical processes such the geological history of the Amazon Basin apparently plays a fundamental role in speciation of the tribe Protiae and consequently in the distribution patterns of its species across Western Amazonia. Fine *et al.* (2005) argued that Miocene-rich sediments constitute the ancestral habitat for most current white-sand specialist species. White-sand habitats were extensive before the Miocene as some evidence suggests and the source of sediments were the Guiana and Brazilian shields. During the Miocene, as the Andean uplift started, the flux of sediments changed dramatically from a westward to an eastward direction, and the much richer sediments originating in the Andes were deposited over most of current Western Amazonia. This geological event produced an incredible edaphic heterogeneity in which parapatric (mosaic sympatry: sensu Mallet *et al.* 2009) speciation promoted by divergent natural selection for habitat specialization would have led to high rates of speciation in the northwestern Amazon forest (Fine *et al.* 2005, 2012; Hoorn *et al.* 2010).

Divergent natural selection should promote habitat specialization and should be involved in diversification processes if populations in contrasting environments (e.g. white sand soils vs. clay soils) evolve a different set of traits that promotes hybrids of the incipient species to have low fitness in both habitats resulting in lineages divergence (Endler 1977; Fine *et al.* 2005).

Despite the fact that we do not have knowledge of any work in Ecuador Amazon supporting this hypothesis, we think there is no reason to doubt the same mechanisms operate across the region (Figure 2.6).

On the other hand, turnover time has been posited as the driver of high diversification rates in Western Amazonia (Baker *et al.* 2014). Using demographic traits data from a plot network in Amazonian forests, the authors found that turnover time improved the estimates of species richness over models of diversification for 51 clades. Faster turnover rates are associated with higher diversification rates if species attain reproductive age faster than expected and populations exhibit short generation times. This could be associated with highly dynamic environments such as the western most part of the Amazon. The region shows a strong influence of Andean uplift, including recent events of deposition of sediments that have originated alluvial terraces, alluvial fans and

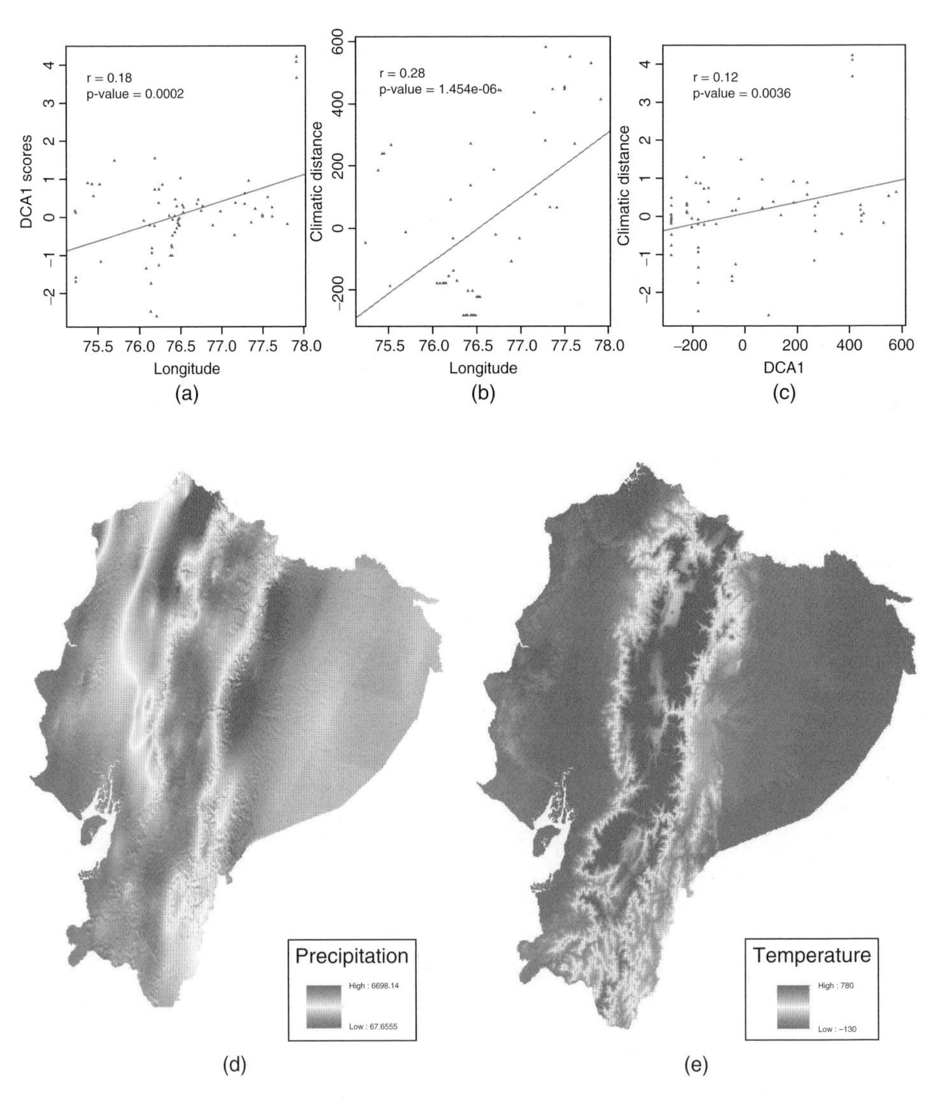

Figure 2.6 Floristic changes are correlated with shifts in climate along a longitudinal gradient in Ecuadorian Amazon forests. (a) Scores of DCA axis 1 against longitude. (b) Climatic distance against longitude; climatic differences were measured as score values of a PCA (principal component analysis) on the basis of monthly mean precipitation values and maximum, minimum and mean temperature values (see Ministerio del Ambiente del Ecuador 2013b). (c) Change in tree species composition as a function of climatic distance. (d) Spatial variation in precipitation across Ecuador. Interpolation of precipitation values was done using the kriging method on the basis of monthly mean precipitation values at 1 km^2 resolution. (e) Spatial variation in temperature across Ecuador. Interpolation of precipitation values were done using the kriging method on the basis of mean, maximum and mean temperature values at 1 km^2 resolution. A longitudinal gradient for temperature and precipitation at landscape scales is clear for (d) and (e).

other geomorphological units with specific underlying soil conditions. These novel edaphic habitats might have promoted high speciation rates in clades with the capability to exploit them efficiently by rapid occupancy and adaptation.

However, turnover times could be assumed to be a by-product of other traits directly involved in reproductive isolation that might promote high diversification rates. Species of the same clade may attain reproductive age faster than close relatives or members of a completely different clade co-existing in sympatry, only if they compete for the same set of potential pollinators. In other words, sympatric linages that compete for the same pollinators may diverge in floral traits promoting increases in diversification rates (Armbruster and Muchala 2008; May and Sargent 2009). This floral isolation may be favored by natural selection after other traits involved in reproductive isolation have evolved.

On the other hand, pollination might reduce extinction rates for rare lineages, insuring that pollen transport between isolated populations could promote an increase in extinction rates if pollination syndromes have evolved through specialization.

In this way increasing turnover times is the result of the natural selection for different pollinators and not the ultimate cause of diversification rates increases.

2.15 Future directions

Regardless of the increasing number of studies that evaluate plant species richness in Western Amazonia, there are still few works concentrating on the Ecuadorean Amazon. Moreover, the lack of works linking taxonomy, phylogeny and traits constrains our ability to determine the potential factors that drive compositional and phylogenetic beta diversity in Amazonian tree communities. Therefore studies combining ecological and evolutionary approaches are necessary to understand the ecological and evolutionary processes acting together in the determination of local assemblages in Ecuador's Amazon forests. For instance, investigating the role of geological formations, soil conditions in the patterns of taxonomic and phylogenetic alpha and beta diversity will require a combination of plot network establishment, geological maps, soil sampling and community phylogenetics to elucidate the importance of these factors.

The inclusion of an evolutionary approach in any analysis of beta diversity is essential in order to improve scientific research-based conservation policies. Because species-centric conservation research takes into consideration a snapshot of the fractal nature of the tree of life, we miss all the information that genealogical relationships between organisms can give us. Evolutionary time plays a fundamental role in the origin and maintenance of biodiversity, because the rate of change in diversity is intrinsically linked to the rates at which a clade or taxa originate and become extinct. For the sake of argument, it is fundamental to consider that regions with extremely high levels of species richness do not necessarily represent areas with high phylogenetic diversity. Furthermore, because phylogenetic beta diversity measures how phylogenetic relatedness changes across environmental and spatial gradients, we can make inferences about the different biogeographical histories of regional species pools with the strong analytical power that phylogenies give to us. Currently, many conservation priority-setting exercises tend to be focused on endemism, whereas many others are based on species information solely and have proved to be a poor predictor of both species richness and threatened species identification. In the light of climate change and human-induced extinctions, it is fundamental to include phylogenetic information in conservation priorities, even more so when the conservation research focus has reached the Noah's Ark dilemma when considering flagship, keystone and umbrella or restricted-range endemic species.

Ultimately, because physiological attributes are intimately related to habitat or ecological space occupancy, the inclusion of ecologically relevant traits is essential to understand both ecological and evolutionary causes of tree species diversity changes across space and environment. Therefore, the future of Western Amazonia's hyper-diverse forests depends on our capacity to develop multi-approach studies in order to understand the processes underlying the community assembly patterns of these systems.

References

Alverson, W.S., Vriesendorp, C., del Campo, Á. *et al.* (eds) (2008) *Ecuador-Perú: Cuyabeno-Güeppí Rapid Biological and Social Inventories Report 20*, The Field Museum, Chicago, IL.

Antonelli, A., Nylander, J.A.A., Persson, C. and San Martín, I. (2009) Tracing the impact of Andean uplift in Neotropical plant evolution. *Proceedings of the National Academy of Sciences of the United States of America*, **106** (24), 9749–9754.

APG III (2009) An update of the Angiosperm Phylogeny Group classification for the orders and families of flowering plants: APG III. *Botanical Journal of the Linnean Society*, **161**, 105–121.

Armbruster, W.S. and Muchhala, N. (2008) Associations between floral specialization and species diversity: cause, effect, or correlation? *Evolutionary Ecology*, **23**, 159–179.

Baker, T.R., Pennington, R.T., Magallon, S. *et al.* (2014) Fast demographic traits promote high diversification rates of Amazonian trees. *Ecology Letters*, **17**, 527–36.

Bass, S.M., Finner, M., Jenkins, C.N. *et al.* (2010) Global conservation significance of Ecuador's Yasuní National Park. *PLoSONE*, **5** (1), 1–22.

Cavender-Bares, K., Kozak, H., Fine, P.V.A. and Kembel, S.W. (2009) The merging of community ecology and phylogenetic biology. *Ecology Letters*, **12**, 693–715.

Cerón, C. and Reyes, C. (2003) Predominio de Burseraceae en 1 ha de bosque colinado, Reserva de Producción Faunística Cuyabeno, Ecuador. *Cinchonia*, **4** (1), 47–60.

Cerón, C.E., Montalvo, C. and Reyes, C.I. (2009) El bosque de tierra firme, moretal, igapo y ripario en la cuenca del río Gueppi, Sucumbíos-Ecuador. *Cinchonia*, **4** (1), 80–109.

Chase, M.W. and Reveal, J.L. (2009) A phylogenetic classification of the land plants to accompany APG III. *Botanical Journal of the Linnean Society*, **161**, 122–127.

Chao, A., Gottelli, N.J., Hsieh, T.C. *et al.* (2014) Rarefaction and extrapolation with Hill numbers: a framework for sampling and estimation in species diversity studies. *Ecological Monographs*, **84** (1), 45–67.

Coley, P. and Kursar, T. (2014) On tropical forests and their pests. *Science*, **343**, 35–36.

Crowther, T.W., Glick, H.B., Covey, K.R. *et al.* (2015) Mapping tree density at global scale. *Nature*, **525**, 201–205.

De Oliveira, A. and Daly, D. (1999) Geographic distribution of tree species occurring in the region of Manaus, Brazil: implications for regional diversity and conservation. *Biodiversity and Conservation*, **8** (9), 1245–1259.

De Oliveira, A. and Mori, S. (1999) A central Amazonia **terra firme** forest. I: High tree species richness on poor soils. *Biodiversity and Conservation*, **8**, 1219–1244.

De Oliveira, S.M. and ter Steege, H. (2013) Floristic overview of the epiphytic bryophytes of *terra firme* forests across the Amazon basin. *Acta Botanica Brasilica*, **27** (2), 347–363 http://dx.doi.org/10.1590/S0102-33062013000200010.

Duivenvoorden, J. (1995) Tree species composition and rain forest-environment relationships in the middle Caquetá area, Colombia, NW Amazonia. *Vegetatio*, **120**, 91–113.

Duivenvoorden, J. (1996) Patterns of tree species richness in rain forests of the middle Caquetá area, Colombia, NW Amazonia. *Biotropica*, **28** (2), 142–158.

Duque, A., Sánchez, M., Cavelier, J. and Duivenvoorden, J.F. (2002) Different floristic patterns of woody understory and canopy plants in Colombian Amazonia. *Journal of Tropical Ecology*, **18**, 499–525.

Endler, J.A. (1977) *Geographic Variation, Speciation and Clines*, Princeton University Press, Princeton, NJ.

Figuereido, F.O., Costa, F.R.C., Nelson, B.W. and Pimentel, T.P. (2014) Validating forest types based on geological and land-form features in central Amazonia. *Journal of Vegetation Science*, **25**, 198–212.

Fine, P.V.A. (2015) Ecological and evolutionary drivers of geographic variation in species diversity. *Annual Review of Ecology, Evolution and Systematics*, **46**, 369–92.

Fine, P.V.A. and Kembel, S. (2011) Phylogenetic community structure and phylogenetic turnover across space and edaphic gradients in western Amazonian tree communities. *Ecography*, **34**, 552–556.

Fine, P.V.A., Mesones, I. and Coley, P.D. (2004) Herbivores promote habitat specialization by trees in Amazonian forests. *Science*, **305**, 663–665.

Fine, P.V.A., Daly, D.C., Villa Muñoz, G. *et al.* (2005) The contribution of edaphic heterogeneity to the evolution and diversity of Burseraceae trees in the western Amazon. *Evolution*, **59**, 1464–1478.

Fine, P.V.A., Garcia Villacorta, R., Pitman, N.C.A. *et al.* (2010) A floristic study of the white sand forests of Peru. *Annals of the Missouri Botanical Garden*, **97**, 283–305.

Fine, P.V.A., Zapata, F., Daly, D. *et al.* (2012) The importance of environmental heterogeneity and spatial distance in generating phylogeographic structure in edaphic specialist and generalist tree species of Protium (Burseraceae) across the Amazon Basin. *Journal of Biogeography*, **40**, 646–661.

Foster, R.B. and Beltran, H. (1997) Vegetation and flora of the eastern slopes of the Cordillera del Cóndor, in *The Cordillera del Cóndor region of Ecuador and Peru: A biological assessment* (eds T.S.K. Awbery and G. Fabregas), Schulenberg, Conservation International, Washington, DC, pp. 44–63.

Funk, W.C., Caminer, M. and Ron, S. (2012) High levels of cryptic species diversity uncovered in Amazonian frogs. *Proc. R. Soc. B*, **279**, 1806–1814.

Guevara, J.E. (2006) *Floristic variation in 23 one hectare plots in Ecuador Amazon* BSc thesis, Pontificia Universidad Catolica del Ecuador.

Higgins, M., Ruokolainen, K., Tuomisto, H. *et al.* (2011) Geological control of floristic composition in Amazonian forests. *Journal of Biogeography*, **38**, 2136–2149.

Hoorn, C. (1996) Miocene deposits in the Amazon foreland basin. Technical comments. *Science*, **273**, 122.

Hoorn, C., Guerrero, J., Sarmiento, G.A. and Lorente, M.A. (1995) Andean tectonics as a cause for changing drainage patterns in Miocene northern South America. *Geology*, **23** (3), 237–240.

Hoorn, C., Wesselingh, F.P., Ter Steege, H. *et al.* (2010) Amazonia through time: Andean uplift, climate change, landscape evolution, and biodiversity. *Science*, **330**, 927–931.

Jablonski, D., Roy, K. and Valentine, J.W. (2006) Out of the tropics: evolutionary dynamics of the latitudinal diversity gradient. *Science*, **314**, 102–106.

Jadan, O. and Aguirre, Z. (2011) Flora de los Tepuyes de la Cuenca Alta del Río Nangaritza, Cordillera del Cóndor, in *Evaluación Ecológica Rápida de la Biodiversidad de los Tepuyes de la Cuenca Alta del Río Nangaritza, Cordillera del Cóndor, Ecuador* (eds J.M. Guayasamin and E. Bonacorso), Conservation International, Quito, Ecuador, pp. 41–48.

Jetz, W. and Fine, P.V.A. (2012) Global gradients in vertebrate diversity predicted by historical area-productivity dynamics and contemporary environment. *PLOS Biol.*, **10** (3), e1001292.

Jorgensen, P.M. and León-Yanez, S. (1999) *Catálogo de plantas vasculares del Ecuador*, Missouri Botanical Press, USA.

Kisel, Y., McInnes, L., Toomey, N.H. and Orme, C.D.L. (2011) How diversification rates and diversity limits combine to create large-scale species-area relationships. *Philos. Trans. R. Soc. B*, **366**, 2514–2525.

Kraft, N., Valencia, R. and Ackerly, D.A. (2008) Functional traits and niche-based tree community assembly in an Amazonian forest. *Science*, **32**, 580–582.

Kreft, H., Koster, N., Kuper, W. *et al.* (2004) Diversity and biogeography of vascular epiphytes in Western Amazonia, Yasuní, Ecuador. *Journal of Biogeography*, **31**, 1463–1476.

Kursar, T., Dexter, K.G., Lokvam, J. *et al.* (2009) The evolution of antiherbivore defenses and their contribution to species co-existence in the tropical tree genus *Inga*. *Proceedings of the National Academy of Sciences*, **106** (43), 18073–18078.

Lujan, K., Roachl, A., Jacobsen, D. *et al.* (2013) Aquatic community structure across an Andes-to-Amazon fluvial gradient. *Journal of Biogeography*, **40**, 1715–1728.

Macía, J.M. and Svenning, J.C. (2005) Oligarchic dominance in western Amazonian plant communities. *Journal of Tropical Ecology*, **21**, 613–626.
Mallet, J.A., Meyer, P.N. and Feder, J.L. (2009) Space, sympatry and speciation. *Journal of Evolutionary Biology*, **22**, 2332–2341.
May, K.M. and Sargent, R.D. (2009) The role of animal pollination in plant speciation: integrating ecology, geography and genetics. *Annual Review of Ecology, Evolution, and Systematics*, **40**, 637–656.
Ministerio del Ambiente del Ecuador (2013a) *Sistema de Clasificación de los Ecosistemas del Ecuador Continental*, Subsecretaría de Patrimonio Natural, Quito.
Ministerio del Ambiente del Ecuador (2013b) *Metodología para la Representación Cartográfica de los Ecosistemas del Ecuador Continental*, Subsecretaría de Patrimonio Natural, Quito.
Mittelbach, G.G., Schemske, D.W., Cornell, H.V. *et al.* (2007) Evolution and the latitudinal diversity gradient: speciation, extinction and biogeography. *Ecology. Letters*, **10**, 315–31.
Myers, N., Mittermeier, R.M., Mittermeier, C.G. *et al.* (2000) Biodiversity hotspots for conservation priorities. *Nature*, **403**, 853–858.
Neill, D. and Ulloa Ulloa, C. (2011) *Adiciones a la Flora del Ecuador: Segundo Suplemento, 2005–2010*, Fundación Jatun Sacha, Quito.
Orme, C.D.L., Davies, R.G., Burgess, M. *et al.* (2005) Global hotspots of species richness are not congruent with endemism or threat. *Nature*, **436**, 1016–1019.
Pennington, R.T. and Dick, C.W. (2010) Diversification of the Amazonian flora and its relation to key geological and environmental events: a molecular perspective, in *Amazonia, Landscape and Species Evolution* (eds C. Hoorn and F.P. Wesselingh), Blackwell Publishing, Oxford, UK, pp. 373–385.
Philips, O., Núñez, P.V., Monteagudo, A. *et al.* (2003) Habitat association among Amazonian tree species: a landscape-scale approach. *Ecology*, **91**, 757–775.
Pitman, N.C.A. (2000) *A large-scale inventory of two Amazonian tree communities* PhD Thesis, Universidad de Duke, Durham, NC.
Pitman, N.C.A., Terborgh, J., Silman, M.R. *et al.* (2001) Dominance and distribution of tree species in upper Amazonian *terra firme* forests. *Ecology*, **82**, 2101–2117.
Pitman, N.C.A., Jorgensen, P.M., Williams, R.S.R. *et al.* (2002) Extinction-rate estimates for a modern neotropical flora. *Conservation Biology*, **16** (5), 1427–1431.
Pitman, N.C.A., Beltrán, H., Foster, R. *et al.* (2003) Flora y vegetación del valle del río Yavarí, in *Perú: Yavarí* Rapid biological inventories report 11 (eds N.C. Pitman, D. Vriesendorp and D. Moskovits), The Field Museum Press, Chicago, IL, pp. 52–59.
Pitman, N.C.A., Mogollón, H., Dávila, N. *et al.* (2008) Tree community change across 700 km of lowland Amazonian forest from the Andean foothills to Brazil. *Biotropica*, **40** (5), 525–654.
Pitman, N.C.A., Guevara Andino, J.E., Aulestia, M. *et al.* (2014) Distribution and abundance of tree species in swamp forest in Amazonian Ecuador. *Ecography*, **37**, 902–915.
Poulsen, D.A., Tuomisto, H. and Balslev, H. (2006) Edaphic and floristic variation within a 1-ha plot of lowland Amazonian rain forest. *Biotropica*, **38** (4), 468–478.

Ricklefs, R.E. (2006) Evolutionary diversification and the origin of the diversity-environment relationship. *Ecology*, **87**, S3–S13.

Rohde, K. (1992) Latitudinal gradients in species diversity: the search for the primary cause. *Oikos*, **65**, 514–27.

Romoleroux, K., Foster, R., Valencia, R. *et al.* (1997) Especies lenosas (dap → 1 cm) encontradas en dos hectareas de un bosque de la Amazonia ecuatoriana, in *Estudios Sobre Diversidad y Ecologia de Plantas* (eds R. Valencia and H. Balslev), Pontificia Universidad Catolica del Ecuador, Quito, Ecuador, pp. 189–215.

Ruokolainen, K., Linna, A. and Tuomisto, H. (1997) Use of Melastomataceae and pteridophytes for revealing phytogeograpic patterns in Amazonian rain forests. *Journal of Tropical Ecology*, **13**, 243–256.

Russo, S.E., Davies, S.J., King, D.A. and Tan, S. (2005) Soil-related performance variation and distributions of tree species in a Bornean rain forest. *Ecology*, **93**, 879–889.

Saunders, T.J. (2008) Geología, hidrología y suelos: procesos y propiedades del paisaje. Ecuador-Perú: Cuyabeno-Güeppí, in *Rapid Biological and Social Inventories* Report 20 (eds W.S.C. Alverson, Á. Vriesendorp, D.K. del Campo *et al.*), The Field Museum, Chicago, IL, pp. 66–75.

Schulte, R.P.O., Lantinga, E.A. and Hawkins, M.J. (2005) A new family of Fisher-curves estimates Fisher's alpha more accurately. *Journal of Theoretical Biology*, **232**, 305–313.

Sierra, R. (2013) *Patrones y factores de deforestación en el Ecuador continental, 1990–2010. Y un acercamiento a los próximos 10 años*, Conservación Internacional Ecuador y Forest Trends, Quito, Ecuador.

ter Steege, H., Pitman, N.C.A., Sabatier, D. *et al.* (2013) Hyper dominance in Amazonian tree flora. *Science*, **342**, 1243092.

Tuomisto, H. and Poulsen, A.D. (1996) Influence of edaphic specialization on pteridophyte distribution in neotropical rain forests. *Journal of Biogeography*, **23**, 283–293.

Tuomisto, H., Poulsen, A.D., Ruokolainen, K. *et al.* (2003a) Linking floristic patterns with soil heterogeneity and satellite imagery in Ecuadorian Amazonia. *Ecological Applications*, **13** (2), 352–371.

Tuomisto, H., Ruokolainen, K. and Yli-Halla, M. (2003b) Dispersal, environment and floristic variation of western Amazonian forests. *Science*, **299**, 241–244.

Valencia, R., Balslev, H. and Pazmiño, G. (1994) High tree alpha-diversity in Amazonian Ecuador. *Biodiversity and Conservation*, **3**, 21–28.

Valencia, R., Foster, R.B., Villa, G. *et al.* (2004) Tree diversity in the Amazon and the contribution of local habitat variation: a large forest plot in eastern Ecuador. *Journal of Ecology*, **92**, 214–229.

Vormisto, J., Svenning, J.C., May, P. and Balslev, H. (2004) Diversity and dominance in palm (Arecaceae) communities in *terra firme* forests in the western Amazon basin. *Journal of Ecology*, **92**, 577–588.

3

Geographical Context of Western Amazonian Forest Use

Risto Kalliola and Sanna Mäki

Abstract

The Western Amazon Region in the Andean foothills and forelands of Colombia, Ecuador, Peru and Bolivia constitutes a vast belt of tropical rainforests in geographically variable settings. The Andean–Amazonian interface characterizes the entire region and makes it distinctive from all other parts of the Amazon Basin. Much of the region's development has been dictated by outside forces from since the colonial period. Many current human arrays in Western Amazonia consequently represent legacies from the colonial and early post-colonial times, when the overall administrative and economic systems with their respective natural resource use practices were established. During independence, parallel domestic developments have occurred in adjacent countries, but with low levels of transnational integration. International agreements and organizations have also contributed to the ways in how environmental policies and decisions are made. At present, the Western Amazonian lands are among the sparsest populated parts in each country, but pressures for their intensified uses are increasing. In this chapter, we integrate environmental and human approaches for an overview of these settings as a retrospective account from the region's early occupation until modern times. Our main conclusion is that although the Andean–Amazonian interface brings the whole area together environmentally, the colonial and post-colonial histories have resulted into low levels of cultural, political and economic coherences within the region as a whole. Moreover, considering Western Amazonia's dynamic and spatially variable environment with its rich biota, current scientific comprehension provides only scarce information to support sustainable land allocation and resource use in the region.

Keywords *Andean foothills; biogeographical theories; forest resources; human development; nature conservation; regional South American economies; social structures; sub-Andean environment; Western Amazon Region*

Forest structure, function and dynamics in Western Amazonia, First Edition.
Edited by Randall W. Myster.

3.1 Introduction

The Western Amazon region includes parts of Venezuela, Colombia, Ecuador, Peru, Bolivia and Brazil. Various criteria can be used for defining its exact boundaries, such as biophysical, administrative, economic and social criteria, but none of them is fully undisputable. The area nevertheless has an interface appearance with aspects of both Andean and Amazonian influences in the form of strong geographical gradients, complex river networks, highly diverse broad-leaved forests, abundant natural resources and different peoples having various kinds of cultural heritages.

Although the Western Amazonian region can be seen as a natural geographical continuum, general scientific writings with this conception are few. Instead, Amazonian research literature commonly focuses some parts only of this region – or more commonly the middle parts of the Amazon Basin, or the Brazilian Amazonia Legal. This chapter consequently tries to make a difference by focusing the Western Amazonian region as a whole and emphasizing the ways and patterns on how its forest resources have been used and governed. Until recent times, these forests have been preserved largely undisturbed due to their difficult access, yet they have for long time also been seen as potential sources of natural resources and wealth. Although their economic uses have followed apparently parallel developments in adjacent countries, detailed level drivers and local development patterns have been disparate.

In the following we will examine the Amazonian regions of Colombia, Ecuador, Peru and Bolivia from both natural and human perspectives as well as their spatial and temporal developments. By using literature sources, maps, statistics and our own research experience in the region, we try to integrate and synthesize the existing scattered research with the following specific objectives:

1. to understand how the sub-Andean environment and different phases of human development have acted as drivers of present administrative, economic and social structures in the region;
2. to assess the present-day practices and patterns of forest use; and
3. to contemplate some current development trends in the region.

3.2 Conditions set by the physical geography

Macro-level landscape characteristics in Western Amazonia vary according to altitude, distance from the Equator, geological structures and climate (Figure 3.1). Altitudinal differences are distinctive near mountains, but decrease in the lowlands where the local topography usually varies to within only a few tens of meters. The Amazon River system as we know it today evolved during the Late Miocene after a phase of fast uplift in the Eastern Cordilleras (Hoorn 1994). The Andean forelands on their eastern side constitute a large sedimentary basin that is bordered by the Guianan highlands to the northeast and the Brazilian highlands to the southeast. The sediments in this basin are mostly of fluvial origin, but in large parts of Colombia and Peru also sediments formed in Miocene

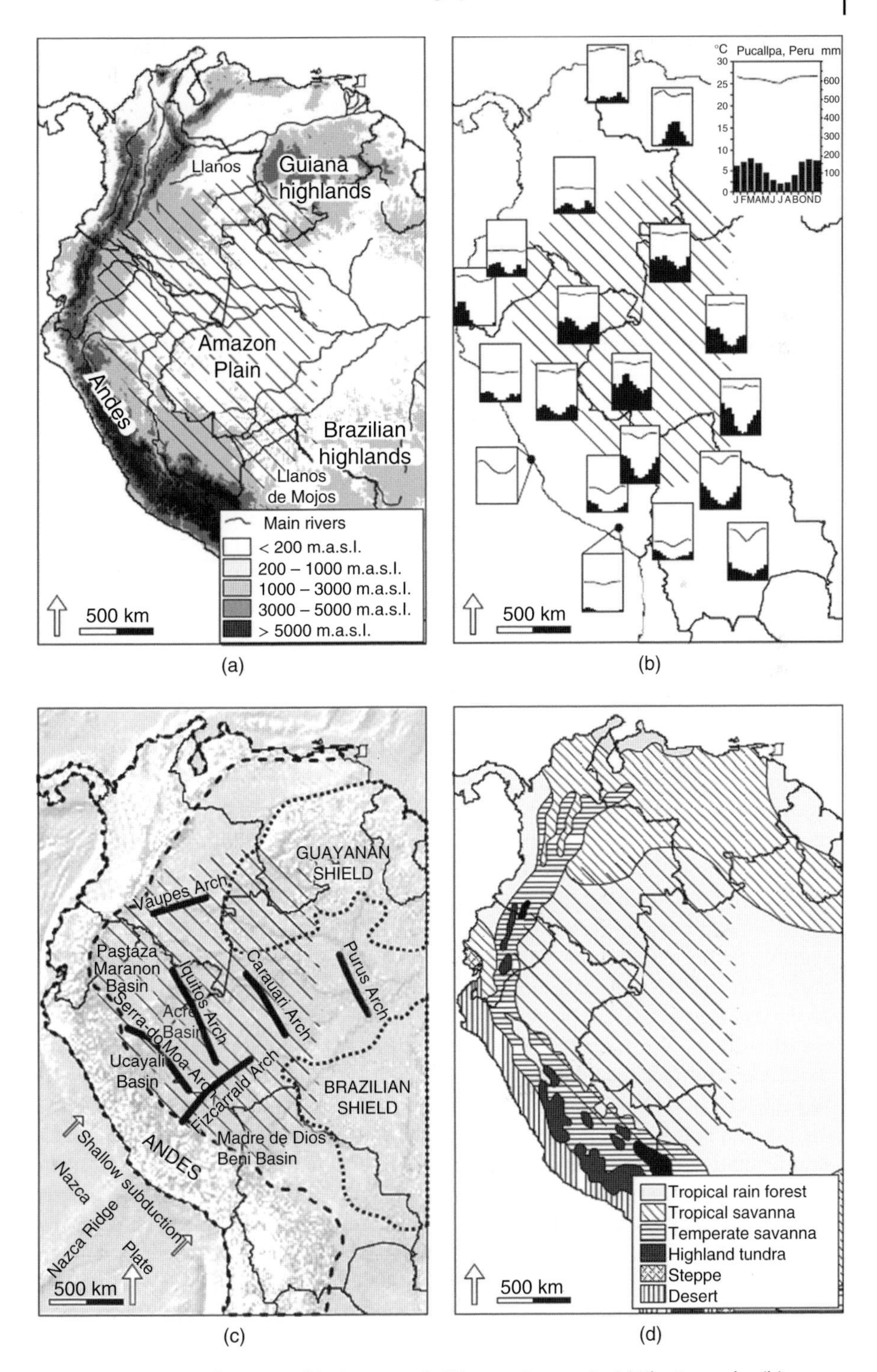

Figure 3.1 Large-scale geographical patterns in Western Amazonia. (a) Physiography. (b) Climate – variations in mean monthly temperature and precipitation (based on: http://www.worldclimate.com/worldclimate/index.htm). (c) Geological formations and dynamics (after Räsänen *et al.* 1987). (d) Vegetation.

amphibious landscapes (Pebas stratigraphies) outcrop widely (Hoorn *et al.* 2010; Räsänen *et al.* 1995). Tectonic compression of the sub-Andean fold and thrust belt induces further differentiation between the geologically distinctive landscape categories. Intra-foreland sub-basins are continually subsiding and being filled with fluvial sediments, and emerging structural heights called arches experience denudation (Irion and Kalliola 2009; Räsänen *et al.* 1992). Tectonic tilting forces some floodplains to migrate laterally and annual inundations are vast, especially in the sub-basins.

River dynamics has great influence on the geographical appearance and regeneration dynamics of floodplain vegetation (Salo *et al.* 1986). White-water rivers, which often have their headwaters in the Andes, are rich in suspended sediments, and when flowing through the lowlands they often have unstable channels and present vast-extending inundations. Large trunk rivers, such as the Marañon, Ucayali, Amazonas and Madre de Dios, have very large floodplains and unstable river channels (Kalliola *et al.* 1992). Black-water streams with their headwaters in the Amazon Basin are more stable, nutrient-poor and acidic (Puhakka *et al.* 1992). Rivers of different kinds are actually so plentiful in Western Amazonia that the entire territory can be regarded as a vast riverscape with three distinctive sub-regions (Toivonen *et al.* 2007): the northern and north-western parts characterized by abundant small- and middle-sized rivers of Andean origin (Ecuador, Colombia), central Western Amazonia in Peru with dynamic trunk rivers running through foreland sub-basins, and the southern part with narrow meandering rivers that are mainly rising within the Amazonian lowlands (Peru, Brazil, Bolivia).

Plant communities in Western Amazonia are fundamentally differentiated by the degree and kind of possible flooding (time, regularity, water chemistry), soil type, micro-relief and human impact (Myster 2009). Part of the region's extraordinarily high species diversity reflects the spatially variable edaphic conditions near the Andean foothills (Tuomisto *et al.* 1995). Up to some 1,500 m a.s.l., Andean forests are floristically similar to lowland Amazonian forests, but at higher altitudes vegetation is different (Gentry *et al.* 1995). White-sand soils with their specialized flora and fauna are scattered in various parts of eastern Colombia, northeastern Peru and eastern Ecuador (Alonso and Whitney 2003). Colombian lowlands have particular vegetation on sandstone plateaus (Arbeláez and Duivenvoorden 2004). Northern Peruvian and Southern Ecuadorian lowland forests are partitioned into large-area units on the basis of less apparent geological formations, which have nevertheless induced edaphic discontinuities and vast extending floristic patterns (Higgins *et al.* 2011). Relationships like these suggest that a significant degree of the present-day forest patterns in the region may result from characteristics of landscape evolution, paleo-environments and dispersal barriers (Tuomisto 2007). However, not all distribution patterns are alike, for example palm distributions are more strongly related to geographical distance than to environmental differences (Vormisto *et al.* 2004). Also areas with species-poor vegetation occur, such as waterlogged swamps (Kalliola *et al.* 1991; Lähteenoja *et al.* 2009) and forests dominated by bamboo (Nelson 1994).

The Western Amazon region has noticeable geographical variations in rainfall, temperature and other climatological properties. In general, the climate is tropical wet near the Equator and tropical wet and dry with some seasonality towards the subtropical zones (WMO/UNESCO 1975). Away from the Equator, broad-leaved forests have in places sharp borders with tropical savannas (in both Colombia and Bolivia). The regional climate system also embodies mechanisms inducing occasional anomalously wet or dry episodes and irregular intrusions of cold air masses, all of which may influence ecosystem functioning (Garreaud 2009; Marengo 1992). The presence of buried charcoal in some areas in Peru and Bolivia furthermore confirms the role of forest fires and pre-Colombian cultures as additional landscape forming disturbances (Bush *et al.* 2007).

3.3 Pre-Colonial human development

Human history in northern South America can be divided into the three distinctive eras: pre-Colonial, Colonial and independence. Many present-day regional features in Amazonia are direct or indirect legacies from those periods (Figure 3.2).

The first people migrated to South America some 15,000 years ago. The earliest inhabitants were supposedly hunter gatherers who occupied a range of different habitats in the mountainous and coastal areas (Fagan 1987). Findings from the Amazon Basin suggest that the first migrations into South America may not have been limited to these areas only. A cave in the Amazon Basin in northern Brazil near the town of Monte Alegre has remains dating to the times of the first migration (Roosevelt 1996).

An important watershed in the South American cultural history has been the invention of ceramics some 7,000 years ago. The first signs of agriculture also began to appear at the same time. The productive forms of economy, agriculture and herding gradually led to regional patterns of production and some sort of trade. The ritual sphere started to develop 5,000–4,000 years ago and triggered the development of high civilizations such as the Chavin and Aymara cultures in the Andes and on the Pacific Coast. The latest and the most widespread pre-Colombian culture was the Inca State (AD 1200–1532), Tawantinsuyu, which extended at its largest from southern Colombia to central Chile and was made of provinces that were attached to the state by personal ties between the provincial leaders and the Inca ruler in Cusco (Pärssinen 1992).

Knowledge about pre-Colombian humans in the Amazonian lowlands is scattered. An obvious reason for this is that artifacts made of organic materials do not preserve well in tropical conditions. The cultural history is consequently mainly evidenced by ceramics, remains of cultivated crops, anthropic soils and burial sites (Gibbons 1990). In recent years, peculiar earthworks known as geoglyphs have been found over a region of *ca.* 65,000 square kilometers in the state of Acre in Brazil after their denudation by modern deforestation (Pärssinen *et al.* 2009). Having geometrical forms such as circles or hexagons, these structures may have had ritual or defense significances. They are consequently strong indications of an advanced and population-rich culture that has lived in the western parts of the Amazon Basin for considerable time (McMichael

YEAR	General history of NE South America	Economy	Western Amazon Region	
15000 BP	Migration to South America; Occupation of Andes and Amazon	Foraging (hunterer-gatherers)		TRADITIONAL CULTURES
10000 BP	Occupation of whole South America		Earliest fingdings of human presence	
8000 BP	Agriculture	Agriculture		
5000–3000 BP	Ritual sphere: emergence of high civilications	Trade	Development of Western Amazon cultural and trade networks	
1200-1532 AD	Inca culture		Incas & other indigenous cultures:interaction and trade relations	
1494	Treaty of Tordesillas	Establishment of the colonial economy	Unexplored Amazon	COLONIAL PERIOD
1500s	Explorations, conquist, settlement	Mineral and agricultural production	First Amazonian explorations 1530s	
1550–	End of conquest; settling of colonies			
1500–1600	Political division		Colonial towns and centres	
1638–1767	Jesuit (and Franciscan) missions	Trade expansion	WAm more tightly under colonial rule	
1800	Liberation started		Scientific explorations since 1700s	
1819–1825	Independence	Export-led economy		INDEPENDENCE
	Wars and boundary disputes, internal unrest	Economic difficulties due to the warfare		
1870	Towards political stability		Rubber boom; opening to the world	
1950s	Nationalism, regionalisation	Import Substitution strategy	Settlement, resource exploitation, agricultural programmes, coca production & trade, timber	
1980	Internal political conflicts, terrorist activities	Economic crisis	Conservation & indigenous rights	
1990	New opening to the world, international agreements and co-operation, different strategies in different countries			

Figure 3.2 A general outline of the civil and economic histories in northwest South America with special reference to the Western Amazon Region.

et al. 2014). It may be interpreted that at the time when these structures were constructed, large areas must have been more or less without tree cover, and after initial clearing, areas were likely kept clear by burning and stump removal (Erickson 2010). These Amazonian people probably domesticated some species, developed cultivation practices, and improved storage and transportation techniques (Denevan 2001). Studies made in the Bolivian Amazon suggest that some of the Amazonian landscape types which at present appear natural may actually be of human origin (Erickson 1995, 2000).

Rivers were used for fishing and transportation, and the fertile floodplain soils supported food production (Denevan 1996). Due to dynamic floodplain conditions, permanent settlements probably preferred upland riverbanks. The Amerindian tribes also organized extensive trade relations among each other (Reeve 1994). Close to the Andes were intermediary groups who formed a trade link between the Andean cultures and Amazonian tribes. There is also some evidence of ritual exchange between the Andes and the Amazonian lands, such as anacondas and caimans brought from the lowlands to Cusco (Lathrap 1973). The parts where the Incas mostly influenced the lowland Amazonia were in the upper reaches of the Beni, Madre de Dios and Ucayali rivers in northern Bolivia and southeast Peru (Pärssinen 1992).

3.4 Colonial era

After the 'discovery' of the New World, Spain and Portugal raised claims for the new territories. The new continent was consequently divided between these countries in the Treaty of Tordesillas 1494 (Wilgus 1967). The Demarcation line followed roughly the meridian of 46° W, but it was later moved several times eastwards by the expansion the Portuguese domain. The Treaty also formed the basis for the first delineation of the Viceroyalty of Peru, which embraced nearly the entire South American continent, except for the Portuguese Brazil, Dutch, French and British settlements on the northern coast (later the Guianas) and the Caribbean Coast of Venezuela (Haring 1973) (Figure 3.3a).

Precious metals and Amazonian forest products were discovered by a number of explorers in the 16th century (Burkholder and Johnson 1990; Smith 1990). The Amazon and Orinoco basins were explored by the expeditions of Francisco de Orellana (1541–1543), Gonzalo Pizarro (1540–1543), Lope de Aguirre (1559–1561) and others. These expeditions were mostly conducted via Quito or Bogota, and the journey descriptions of the early explorers and missionaries were later used as proof that the areas close to the Amazon river should belong to the Ecuadorian territory rather than Peru (Gómez 1996). Explorations from the 1530s onwards also included several attempts to find the mystical town of "El Dorado" with its rumored treasures. However, many explorers failed to penetrate deep into the lowlands or return back due to obstructions such as diseases, environmental harshness, lack of maps and hostile native groups.

The early journeys by Francisco de Orellana provided the first written documentation of the Amazonian territory and settlements along the river. Accordingly, the Amazon Region at that time was not an "empty land", but constituted many native people's settlements with complex societies (Gibbons 1990). During the following centuries, knowledge about the Amazon region increased by later explorers and scientists from all over the world (Smith 1990). On the threshold of the 19th century, Alexander von Humboldt and his colleague Aimé Bonpland discovered a river connection between the Amazon

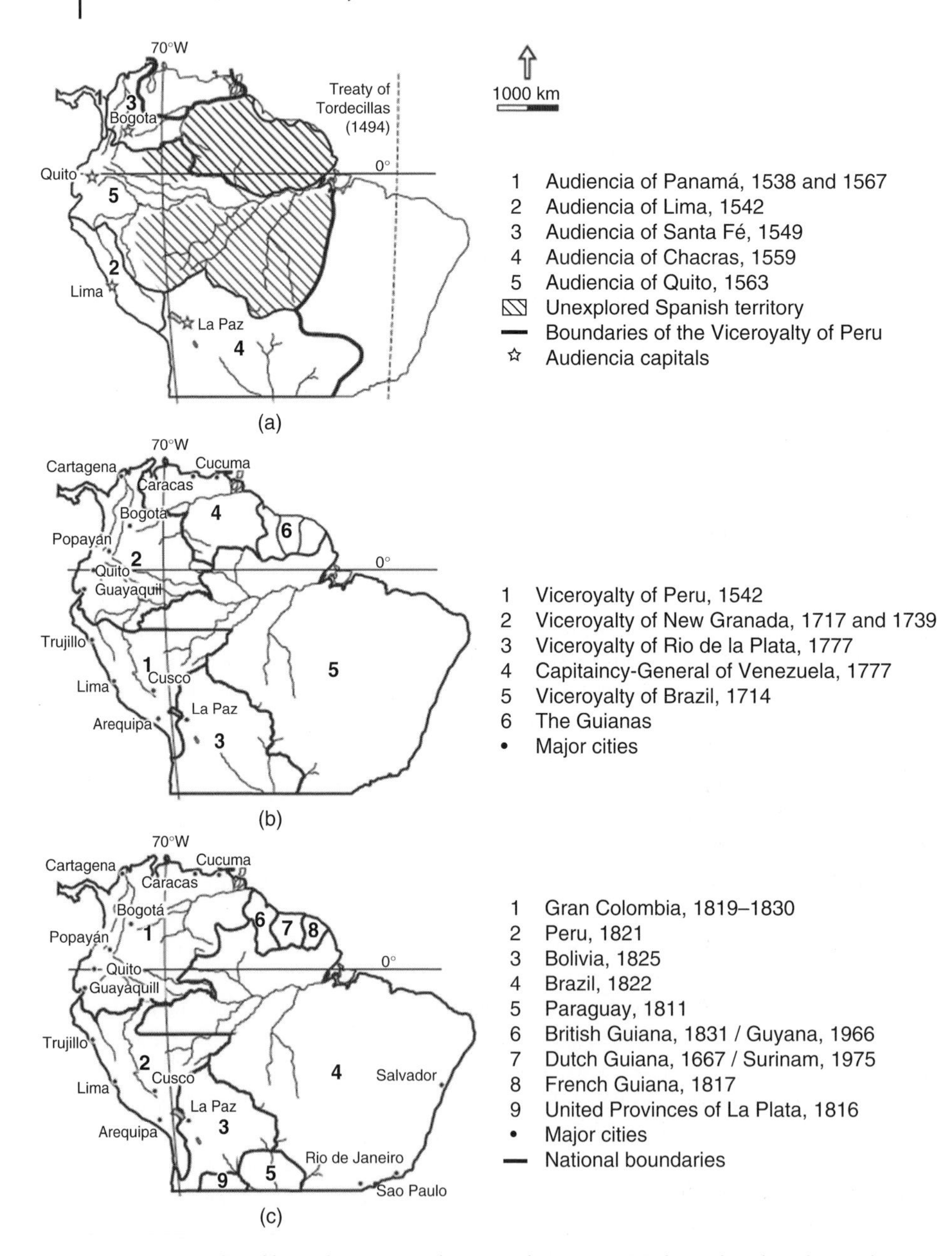

Figure 3.3 Political boundaries in northern South America. (a) The early colonial period *ca.* 1650. (b) The late colonial period *ca.* 1800. (c) After the liberation *ca.* 1830 (drawn according to Lombardi *et al.* 1983).

and the Orinoco systems. Their writings about the geography, geology, ecology, meteorology, archaeology and anthropology of northern South America are still important documents today. They also inspired an increasing number of naturalists and ethnographers to explore the Amazon. As a result, economically valuable tropical products such as medicinal plants and extracts, wild rubber, oils and fibers were discovered.

After the Inca State was conquered, the colonial rule was first established by armed dominance and then by papal right (Sempat Assadourian 1992). The first phase of the colonial rule (1570–1580) consisted of shipments of silver and other valuable minerals to European cities. In the second phase (1570–1620), exports extended to agricultural crops, timber and other forest products. At the same time, the colonial rule gradually expanded into lowland Amazonia. Trade was only allowed between the colony and the Spanish Empire and not between or within the colonies. The system became highly corrupt and bureaucratic, which hampered the growth of regional South American economies.

The Viceroyalty of Peru remained as the only colony throughout the 16th and 17th centuries (Haring 1973). It was soon deemed too large for administrative purposes, and so the Crown divided it into units called *audiencias,* each holding a Royal tribune (Lynch 1992). These were further divided into districts and subdivided into municipalities, the organization forming a pyramidal structure culminating in a centralized authority held by the Crown in Spain. The Audiencia of Lima was founded in 1544 and nearly all of Spanish South America was for a while under its jurisdiction (Haring 1973). The foundation of the Audiencia of Santa Fe de Bogotá in 1549 (changed into the Captaincy-General of New Granada in 1563) meant the end for the period of conquest and a beginning for a more settled period. A land division system called *encomienda* was created during the colonial period and it has a far-reaching legacy even in the present-day land ownership structures in Amazonia (Burkholder and Johnson 1990). In later phases, encomiendas were changed into large estates called *latifundios* and *haciendas.* The patrons of these lands utilized the local natural resources and labor in a feudal manner.

During the early colonial governance, the Western Amazon region remained unexplored, except for the main channel of the Amazon river and a few larger tributaries near the Andes (Lombardi *et al.* 1983). In the latter part of the 16th century and the first half of the 17th century, however, most of the region was already under Spanish rule and partly occupied by the soldiers of the Crown, and also by Jesuit and Franciscan missionaries. Spanish colonization usually advanced as a three-step process: founding towns as nodes to the Andean base, establishing encomiendas close to the towns and along the main rivers, and converting the native peoples (Reeve 1994). The most important towns were founded near the Andean foothills, and the main rivers served as corridors for their administrative, military and trading activites (De Boer 1981).

Between the years 1638 and 1767, the Jesuit missions had considerable impact on the regional economy and demography. The intermediary groups closest to the Andes first entered into trade relations with the Europeans. The more eastern groups also took part in the trade, but their former dominance was diminished by the Spanish occupation and military activities from the west, as well as the threat of being captured as slaves by the Portuguese from the east. The native groups consequently had to balance between the demands of the encomiendas, the military and slave raid threats and their traditional livelihood. Many groups welcomed the Jesuit and the Franciscan missions, as they protected them from military violence, yet many groups were rather "missionized" with the help of the military force (Reeve 1994). The most important traded goods from Europe into the mission settlements were manufactured products such as iron axes, knives and machetes. The indigenous peoples in turn traded captured slaves, blowgun-dart poison, canoes, food and woven cloth. Since the supply of European goods from Quito was insecure, the Jesuits started to explore possibilities for exploiting the forest products such as laurel wax, gums and resins, oils and local handicrafts, which were useful to the colonial economy.

During the Jesuit era, growing violence between the native peoples and the Europeans was manifested by attacks on the colonial towns and missions, and reprisals by the Spanish military forces. Some native groups allied with the Spanish to avoid detrimental confrontations. There were also conflicts among the indigenous groups, the most powerful ones exploiting the smaller ones. These violent acts together with the Old World diseases reduced the number and power of all native groups. However, as settlements were rather dispersed in the forest, the deaths caused by epidemics were in general less severe than in the densely populated Andes (Cleary 2001). The Jesuit missions were mostly concluded after 1767 due to the Crown's concern about the Jesuit's control of trade (De Boer 1981). Commerce was then shifted to the hands of white colonial traders who diminished the role of native traders. This destroyed some of the regional unifying dynamics that had already started to grow in Western Amazonia (Reeve 1994). Some native groups still maintained trade relations with the Spanish, while many others escaped into increasingly remote areas, which elucidate the myth of an uninhabited and pristine Amazonia before the European immigration (Burkholder and Johnson 1990).

In the 18th century, Spanish South America was divided into three viceroyalties: Peru, New Granada (from which the Captaincy General of Venezuela was separated in 1777), and Rio de La Plata (or Buenos Aires) (Haring 1973) (Figure 3.3b). Legacies from these territorial divisions can still be seen in the boundaries of the present-day independent countries. Western Amazonia was shared between the New Granada and the Captaincy General of Venezuela (present-day Colombia, Ecuador and Venezuela), Peru and Rio de La Plata. Noteworthy, the Spanish geopolitical strategies and administrative systems in Amazonia differed from those of Portugal. Spanish administration split the vast region into sub-regions with different levels of autonomy. As a result, the Western Amazon region became divided and disconnected. This differs from the case of Portugal, which expanded its South American holdings by occupying new border areas and using the region's navigable river network to link the Amazonian areas with the rest of the country.

3.5 Liberation and forming of nations

The liberation of northern South America from the power of the Spanish Crown started in the early 1800s (Lombardi *et al.* 1983). The external reasons for the disintegration of the empire were mainly related to weak and corrupt administration, economic disagreements and expensive warfare between Spain and France (Haring 1973). The trade with the colonies also fell increasingly into the hands of intermediate traders, which created dissatisfaction. Further motivating factors included the frustration against the rules set by the Crown. Also the *criollo* councils formed by the colonial elites strove for independence by either political or military means (Burkholder and Johnson 1990).

The most famous liberators of the northern South America were Simon Bolivar, José de San Martín and Antonio José de Sucre. Gran Colombia uniting Colombia, Venezuela and Ecuador was founded in 1819 when Bolivar fought for the independence of these countries. Peru was liberated in 1821 and Bolivia in 1825 (Figure 3.3c). In 1830, Gran Colombia dissolved into Venezuela, Colombia (the first Republic of New Granada from which Panama separated in 1903) and Ecuador (Figure 3.4). Several wars and other boundary disputes finally shaped the present-day territories of the Western Amazon countries (Figure 3.5).

After independence, political and economic instabilities continued for decades. Colombia first turned into the Federation of the United States of Colombia in 1863, which was abolished with a new Constitution in 1886, dividing Colombia into departments with some local autonomy. Ecuador changed its constitutions 8 times during the 19th century and 11 times more in the following century (Bakewell 1997). Peru had a succession of generals in rule and several external conflicts during the 19th and 20th centuries. It had a short confederation with Bolivia (1837–1839), fought against Spain (1864–1871) and went to war with Chile (1879–1884). In 1867 and 1903, Bolivia ceded a substantial part of its possessions in the eastern lowlands, rich in rubber, to Brazil. Defeated in a war with Paraguay (1928–1935), Bolivia also lost most of Gran Chaco. Consequently, the present extent of Bolivia is less than a half of its territory than it was during early independence times.

Ecuadorian Amazonia was defined in the Protocol of Mosquera-Pedemonte (1830) as including the areas between the Caquetá, Marañon and the Amazon

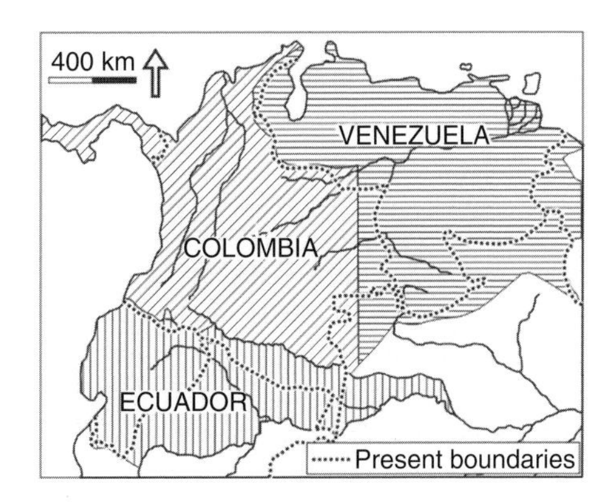

Figure 3.4 The dissolution of independent Gran Colombia in 1830 (based on Lombardi *et al.* 1983).

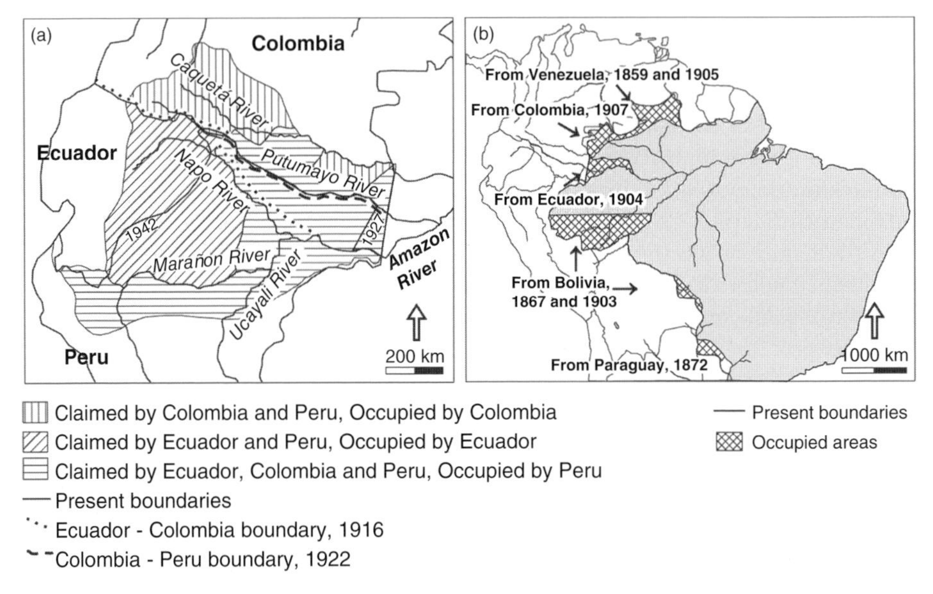

Figure 3.5 (a) Boundary disputes between Colombia, Ecuador and Peru (1830–1942). (b) Brazilian territorial expansion (based on Lombardi *et al.* 1983).

rivers (Gómez 1996). In the treaties with Brazil (1904) and Colombia (1916), Ecuador lost an extensive part of its northern Amazonia. In the treaty of 1922, Peru had ceded the northern part of the occupied territory to Colombia, and in 1927 Colombia gained access to the Amazon river. Peru and Ecuador had long-lasting border conflicts, some perhaps having their deep roots in the territorial organization of the Inca State. Peru occupied increasing amounts of Ecuadorian territories until a short war between the countries ended the dispute in favor of Peru in 1942 (Gómez 1996). Disagreements still continued and the newest peace treaty between these countries was actually established in 1998 with expectations of binational benefits (Ministerio de Relaciones Exteriores del Perú 2002).

3.6 World market integration and changing political regimes

Early independence encouraged free market economies but falling exports due to independence wars followed in the first few decades (Bakewell 1997). The second half of the 19th and the first quarter of the 20th centuries showed export-led growth by world market demand for primary goods in Europe and North America. Booming exports followed for wild rubber and some other forest products (Coomes 1995). The growth was however vulnerable to changes in supply, demand and markets (Larrea and North 1997). The Western Amazonian countries continued to have a status of economic colonies of their export partners. Up until the mid-1800s, Britain had the dominant role in South American trade, but by the early 1900s they were overtaken by the United States.

During the rubber boom (1860–1920), Amazonia dominated the production and international trade of wild rubber. The fluvial network in the region provided important transportation routes to the Atlantic Ocean, and increased trafficking enhanced the region's internal dynamics in an east–west direction. The business was operated by foreign-owned export houses with their returns in the region mainly invested in showy buildings and in rubber business infrastructures. The port cities of Belém, Manaus and Iquitos grew rapidly, but there was little interest to other developments in the region (Higbee 1951). A sharp decline in the rubber economy took place around the 1920s when the plantation production in Asia started to flourish. There were also internal reasons for the decline, such as ineffective monopolies, spatially scattered production areas, problems with labor and high transaction costs. Important for today's Amazonia, some economic and administrative structures from the Rubber Boom era outlived and are still present as cultural legacies in today's Amazonia (Barham and Coomes 1996). Iquitos, for example, grew to become the foremost economic center in the Western Amazon region.

Countermeasure to the export driven economic policies, countries moved towards tighter internal integration and an Import Substitution Strategy (ISI) was implemented to reduce external dependence (Cardoso and Fishbow 1992; French-Davis *et al.* 1998). The original goal was to diversify national economies by using exports to finance industrialization (Philip 2001), but the strategy eventually led to decreasing export volumes as state intervention and incentives generated anti-export bias. Since the economies became increasingly closed, they also missed out on opportunities of the post-war boom in international trade after World War II. Amazonian forests remained an underused resource, although there were expectations to get them better integrated into the national economies (Brown *et al.* 1994).

In the latter part of the 20th century, international organizations started to support Amazonian colonization (Eastwood and Pollard 1992). Road construction in the Andean foreland zone started in the 1940s and was particularly intensive in the late 1950s and 1960s when governments implemented their rural development programs (Murphy *et al.* 1997). The construction of the north–south directed Carretera Marginal was in particular expected to raise economic dynamics in the Andean foothill zone (Brown *et al.* 1994). However, due to insufficient planning and environmental knowledge, expectations of economic revenue were unrealistic. It was also heavily criticized for its expected negative socio-economic and environmental outputs. Progress mainly took place in urban centers, which enhanced rural-urban migration (Cardoso and Fishbow 1992). Also land speculation, lack of local knowledge, environmental harshness and lower than expected marketable surpluses caused severe ecological and economic setbacks (Pichón 1996; Rudel 1983; Schuurman 1979). Towards the end of the century, international concern about the consequences of deforestation resulted in the withdrawal of international funding for road construction projects (Eastwood and Pollard 1992).

In the 1960s, exploitation of hydrocarbon resources started in many parts of Western Amazonia. The resulting economies grew to be essential for the indebted Amazonian countries, but increasing oil production also caused environmental and health risks and gave rise to land use and human right conflicts. For example, in Ecuador, oil production was introduced into both conservation areas and indigenous territories with many kinds of undesired side effects, such as toxic wastes and oil released into pristine nature causing widespread damage to the environment and for the people (San Sebastián and Hurtig 2005). The exploration and production of hydrocarbon resources also involved the building of terrestrial transportation infrastructures as straight transects with cleared forest to cross uninhabited lands. These lineaments have helped colonists to occupy new land areas and to exploit their forest resources.

During the 1960s, countries started to change their economic policies back from strict protectionism. Governments provided export subsidies for non-traditional products such as petroleum, natural gas and timber that involved the flow of large private capital by the foreign investors (Cardoso and Fishbow 1992). Economies were also pursued by the economic integration of Colombia, Ecuador, Peru, Bolivia, Chile and later Venezuela by the Andean Pact in 1969 (French-Davis *et al.* 1998; Khazeh and Clark 1990). In the 1996 reform, the name was changed to the Andean Community of Nations and considerable restructuring of the Pact took place (West 2001). During the 1990s, Bolivia became associated with the Southern Common Market (Mercosur), Colombia pursued tighter economic integration with Ecuador, Venezuela and Mexico, Ecuador strove for increased integration with Colombia and Brazil, and Peru has been diversifying its trading partners whilst also maintaining active integration with the original Pact members (West 2001).

In the late 1900s, Western Amazonian countries experienced internal political conflicts and economic crises such as guerilla and terrorist activity, drug production and traffic, and conflicts between indigenous peoples and colonists (e.g. Franco and Godoy 1992; Meertens and Segura-Escobar 1996; Steinberg 2000). Despite constitutional democracies, all countries also went through periods of autocracy and military dictatorships, as well as periods under the rule of populistic presidents and corrupt governments. These problems resulted in frequent shifts in priorities and left many projects unfinished. Severe debt crisis resulted in vicious circles of hyperinflation. In Amazonia, political and economic fluctuations manifested as conflicts between different interest groups, unsustainable socio-economic structures, illicit economies and non-sustainable resource use. Towards the end of the 20th century, increased inter-regional and international cooperation providing tools for sustainable development started to raise more positive expectations. The early 21st century has likewise reinforced the national and societal stabilities in all countries.

Some typical occurrences during the colonization period can be illuminated by using Eastern Ecuador (Ecuadorian Oriente) as a microcosm where these developments were particularly ample (Figure 3.6). Development in Colombian Amazonia suffered from long periods of rural insecurity and violence related to revolutionary activities and the drug trade (Schuurman 1979). In Peru and

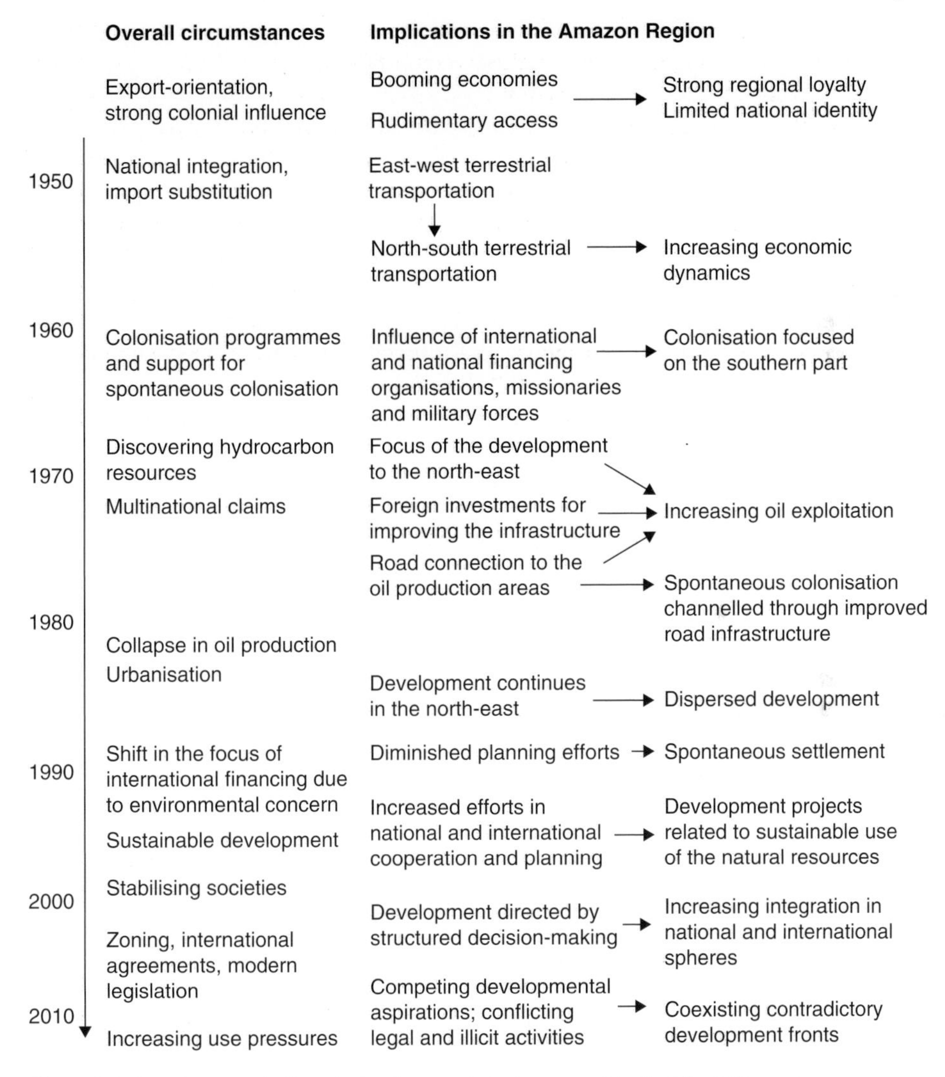

Figure 3.6 Development in Ecuadorian Amazonia in the second half of the 20th century (Eastwood and Pollard 1992).

Bolivia, national decentralization and agrarian reforms were important catalysts for rural Amazonian development (Coomes 1996; Thiele 1995).

3.7 Characteristics of the present forest use

The most important contemporary concern about Amazonian development is the large-scale forest clearance that is widely practised for agricultural production and other purposes (Skole and Tucker 1993; Sun *et al.* 2015). Deforestation makes species vulnerable, destroys crucial forest services, decreases the carbon stored in living vegetation and threatens the surrounding forests through edge effects

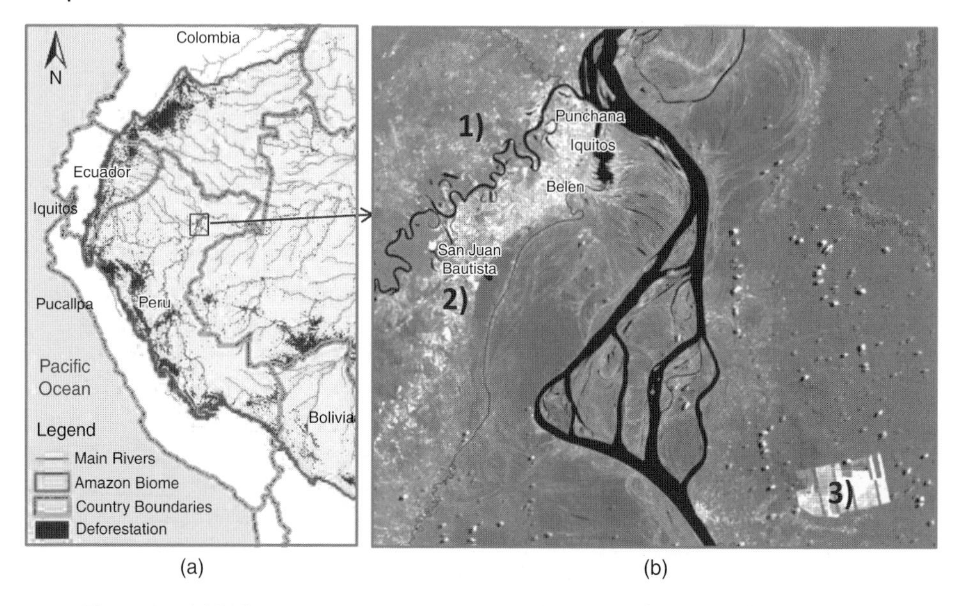

Figure 3.7 (a) Deforestation in Western Amazonia (courtesy of http://wwf.panda.org/wwf_news/?208511/Keeping-an-eye-on-deforestation). (b) Aerial view showing deforestation patterns near Iquitos, Peru (numbers explained in the text) [(b) image from http://www.esri.com/landing-pages/software/landsat/unlock-earthssecrets].

(Nepstad *et al.* 1999). Deforestation is mostly concentrated along the Andean foothill zone, but smaller concentrations of cleared land are also scattered widely across the lowland region (Figure 3.7).

The latter areas can be roughly categorized into three kinds, each corresponding to a particular socioeconomic and ecological setting:

1. spontaneous-looking, gradually expanded and patchy field and fallow systems by rural peoples along and near rivers;
2. systematically expanding deforestation fronts close to population centers and along roads which attract colonization, and
3. large-scale forest clearings raised by big capital for meat or cash-crop production.

For example, a several kilometers-long plantation was carved in early 2010 near Iquitos in an almost intact forest landscape (marked with number 3 in Figure 3.7; World Resources Institute 2015). Land conversions at equal or even larger scales are also made for oil palm, soya and especially cattle ranching. Widespread large-block deforestation is particularly abundant near the city of Cobija at the Bolivian border with Brazil.

Selective logging induces diffuse canopy thinning with many kinds of ecologically significant consequences (Foley *et al.* 2007; Shearman *et al.* 2012). Since the extraction of valuable timber is the most common forestry practice in the region, such hidden consequences can be wide, extending along rivers, roads and other areas of easy access. In the same manner, there are depleted stocks of wild game and parrots in easily accessible lands due to intensive hunting (Salo *et al.* 2014). These kinds of wild species harvest settings resemble non-renewable resource mining as they inevitably lead to declining resource availability unless management efforts to the contrary are successful. Depleted forest resources lower the value of the remaining forest and add the risk of denudation. Such development commonly takes place in front of penetrating roads before larger-scale land conversion follows (Mäki *et al.* 2001). Thoughtful planning could lower the side damage of big tree felling and log transport, but in remote areas active measures supporting reduced impact logging may not be prioritized. Forest certification potentially encourages sustainable forest management by securing a greater income from the harvested products. According to empirical research in Ecuador and Bolivia, the feasibility of this mechanism however varies between different regions (Ebeling and Yasué 2009).

Forest legislation has been modernized in many Western Amazonian countries by the shift into the 21st century. For example the Peruvian forestry law identifies different kinds of forests: permanent production forests, forests for future use, forests on protected lands, protected areas, forests on native and communal lands, and local forests (Salo and Toivonen 2009). As production forests are mostly state-owned, concession systems are used to lease lands for harvesting timber and/or other renewable forest products (Walker and Smith 1993). Concessions are long-term commitments (usually 40 years) requiring proficient forest inventories and plans for future operative and management activities. In the Peruvian region of Loreto, public tendering of forest concessions was executed in 2002 but resulted in less demand than offered (Salo *et al.* 2011). This experience suggests that the forestry sector in this region was not economically appealing at that time or the way how the concession system was applied was not fully attractive. Also the pricing policy by forest authorities is reported to encourage logging made outside rather than within timber concessions (Giudice *et al.* 2012). Illicit logging is indeed a common problem in the Western Amazon and difficult to uproot due to the remoteness of the harvest areas, often unclear administrative circumstances and inefficient control.

Historically, market-driven Amazonian timber extraction has addressed high-value timber species such as *Swietenia macrophylla, Cedrela odorata, Ceiba pentandra, Maquira coreaceae, Virola surinamensis* and *Carapa guianensis* (Pinedo-Vasquez *et al.* 2001). Due to increased international concern the export of these and tens of other tropical timber species from Western Amazonia have diminished, whilst the contemporary logging pressure mainly comes from domestic demand. Country-level export statistics show that Ecuador is a major exporter of roundwood (Table 3.1), despite this export being banned for other purposes than scientific and experimental uses (Sustainable forest products 2015). Also other Western Amazonian countries have restrictions on log exports from natural forests. In sawn wood, Peru shows a clear rise between 1999 and 2013. Compared to Brazil with its sawn-wood exports in 2013 as 1,206,180 m^3, Peru is however a relatively small actor in the world market.

Floodplains of suspension-rich white-water rivers can provide high yields of harvestable timber and withstand relatively strong management interventions (Nebel *et al.* 2001). Forestry on these lands could consequently be sustainable,

Table 3.1 Primary wood product trade export statistics in 1999 and 2013 at country level. (source: ITTO Annual review statistics database http://www.itto.int/annual_review_output/?mode=searchdata).

	Industrial roundwood exports (1,000 m^3)		Sawnwood exports (1,000 m^3)	
	1999	*2013*	*1999*	*2013*
Bolivia	3.17	10.25	41.88	31.63
Colombia	16.73	42.80	9.08	3.95
Ecuador	141.21	170.34	20.61	134.06
Peru	0	0.30	74.00	145.74

taken that resource depletion by overuse is avoided. In non-flooded lands, small-scale logging may mimic natural gap dynamics but the regenerative success of different timber species varies, which calls for skilled and dedicated management practices (Karsten *et al.* 2014). In 1980, an initiative to promote forest plantations in the Amazon region (*Canon de Reforestación en Amazonia Peruana*) was created and it operated until 2000 with about 100,000 hectares of forest-established plantations under this scheme (ITTO 2009). Despite measures to activate the forestry sector, lacking national forest research programs and low numbers of research staff with insufficient resources are clear weaknesses in all Western Amazonian countries. The training of forestry professionals can consequently only build on weak support by scientific proof and experimental work. This hindrance easily leaves the practical managers without solid professional support and lets harvest dominate the sector. Problems in management planning may already arise from the broad variety of environmental settings in the region, which calls for tailored management schemes for different kinds of forests.

Timber and non-timber forest products (NTFPs) are also used and managed on a minor scale within smallholders' diversified livelihood strategies (Mejia *et al.* 2015; Pinedo-Vasquez *et al.* 2001). NTFP harvest is ecologically reasonable and its economic potential may theoretically overrun the short-term benefits of land conversion (Vasquez and Gentry 1989). However, the thesis of "conservation through NTFP commercialization" may underestimate such practical challenges as scarce and uneven distribution of the harvested resources, transport costs, processing needs, local socio-economic settings, and a variety of different market forces (Arnold and Pérez 2001). The production and export of Brazil nuts is a prime example of NTFP-based economy in Peru and Bolivia but even this livelihood is not without problems of viability and poverty (Kalliola 2011). On the other hand, forests should not be valued upon their potential for pure economic return only. Forest services are vital for the local peoples in the form of nutrition, health, housing, clothing and cultural values. This notion is emphasized in extractive reserves that have been established in some areas to sustain subsistence economies. Experiences from their formation in the Amazonian region have however varied, depending on a variety of regional socio-ecological settings (Zeidemann *et al.* 2014). Also the principles of indigenous forest use and management with no involved monetary mechanisms provide practical models of long-term forest usage without resource depletion (Becker and León 2000). In a broader view, Amazonia's forests provide ecosystem services of world-class importance. The accountability of NTFP economies could consequently be reinforced by subsidy mechanisms based on global benefits through, for example, the mechanism of REDD (Reducing Emissions from Deforestation and forest Degradation). The practical implementation of such subsidy measures requires organizational skills and can involve notable transaction costs, both of which are challenging for smallholders (Pokorny *et al.* 2013).

International environmental conventions, such as the Convention on International Trade in Endangered Species of Wild Fauna and Flora (CITES, signed in 1973) and the Convention on Biological Diversity (CBD, signed in 1992), also define conditions for using the region's biological resources. They oblige the signatory countries to take action in their respective fields and furthermore they have turned the world's eyes to see changes in the Amazonian forests. The extraordinarily rich floras and faunas in these forests are seen as biological resources which may bring in wealth, taken that they are used wisely, based upon strategic planning with evidence-based decision-making (BIODAMAZ 2001). However, the raised expectations about the economic potential of the region's extraordinary biodiversity resources do not easily come through (Salo *et al.* 2014). By far the most promising and influential way of using the rich nature *per se* is ecotourism. It has benefited from Amazonia's positive environmental fame and the consequent economic return has augmented local support for nature conservation. However, expectations about the economic return from the ecotourism industry may be overly optimistic. Currently there are only a few quantitative studies of the actual size and profitability of this sector in Western Amazonia (Kirkby *et al.* 2011).

Nature conservation areas in Amazonia have mainly been established after the 1970s, with their delineations largely based on general biogeographic theories and assumed biodiversity representativeness (Jorge Pádua *et al.* 1974; Peres and Terborgh 1995). As none of the used biogeographical theories or datasets define the biological values reliably, the present conservation network in Amazonia may be ecologically biased (Schulman *et al.* 2007). Nonetheless, protected areas are large and plentiful as compared to the forest areas in temperate and boreal zones (Figure 3.8). Dedicating extensive areas for nature protection has been possible for the low pressures from competing land uses at the time when the conservation decisions were made. Nature protection has also been facilitated and strongly pushed forward by external forces such as international organizations and activists. Increasing population and land use pressures may in the future drive countries to reconsider some earlier conservation decisions. For example, two-thirds of Peruvian and Ecuadorian lowlands are covered by oil and gas concessions, with an overlap of 17% with protected areas (Figure 3.8; Finer and Orta-Martínez 2010). Also illegal timber and NTFP extraction in conserved lands are creating conflicts with the reserve authorities (e.g. Messina *et al.* 2006). Despite problems, conservation seems to work: only 2% of forest disturbances between 1999 and 2005 in Peruvian lowland Amazonia occurred within the boundaries of natural protected areas (Oliveira *et al.* 2007). As to

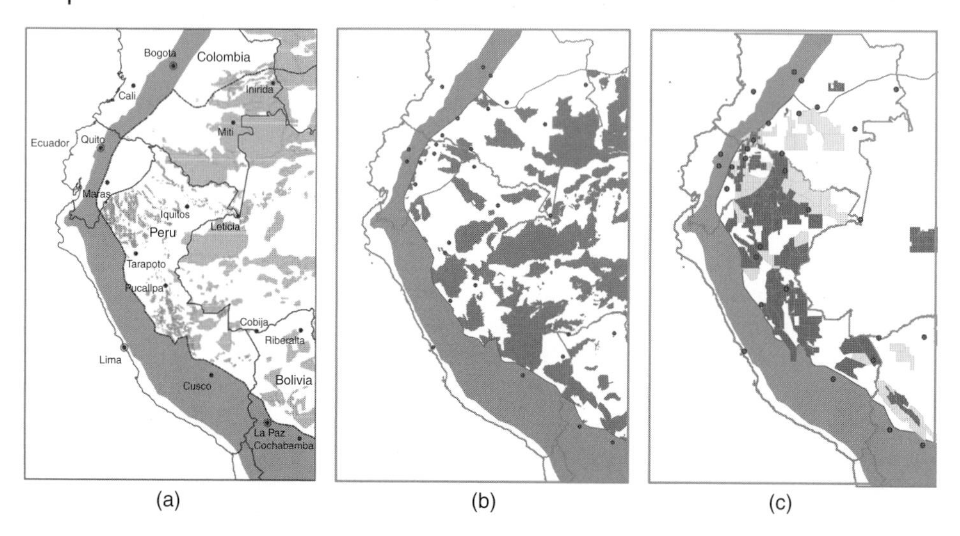

Figure 3.8 Territorial reservations in Western Amazonia. (a) Indigenous territories. (b) Nature protection areas. (c) Hydrocarbon production areas in different stages of accomplishment (dark grey: leased; light grey: not yet leased) (data from https://raisg.socioambiental.org/).

overlaps between nature conservation areas and indigenous people's reserves, they are seldom conflict loaded (Carneiro da Cunha and de Almeida 2000).

Mining is a big problem in some parts of the region; in particular gold mining in the Peruvian department of Madre de Dios has attracted thousands of legal and illegal miners to work in river floodplains. During the monitoring period from 1999 to 2012, Asner *et al.* (2013) observed 50,000 ha of new mining areas of different sizes along several sub-Andean rivers in the region, at places forming large mining landscapes. These activities pose a threat to the region's forests and their wildlife, pollute water resources with mercury and other substances and increase sediment loads in rivers.

3.8 Present population and regional integration

Western Amazonian countries are administratively divided into departments (Peru and Bolivia), states (Colombia) or provinces (Ecuador). They are in general large in size but with small populations and very low population densities (Figure 3.9). Much of the region is a kind of backyard in their respective countries where the main population and economic activities are in the Andean and coastal areas. Western Amazonian lowlands however have distinct Andean and Pacific cultural orientation. This characteristic, too, separates them from Brazilian Amazonia where many cultural aspects have continental and/or Atlantic flavor.

The total population in the Western Amazonian departments/states/provinces is about 4.4 million (Table 3.2), out of which 16% are in Colombia, 17% in Ecuador, 58% in Peru and 9% in Bolivia. Population growth has in general been highest

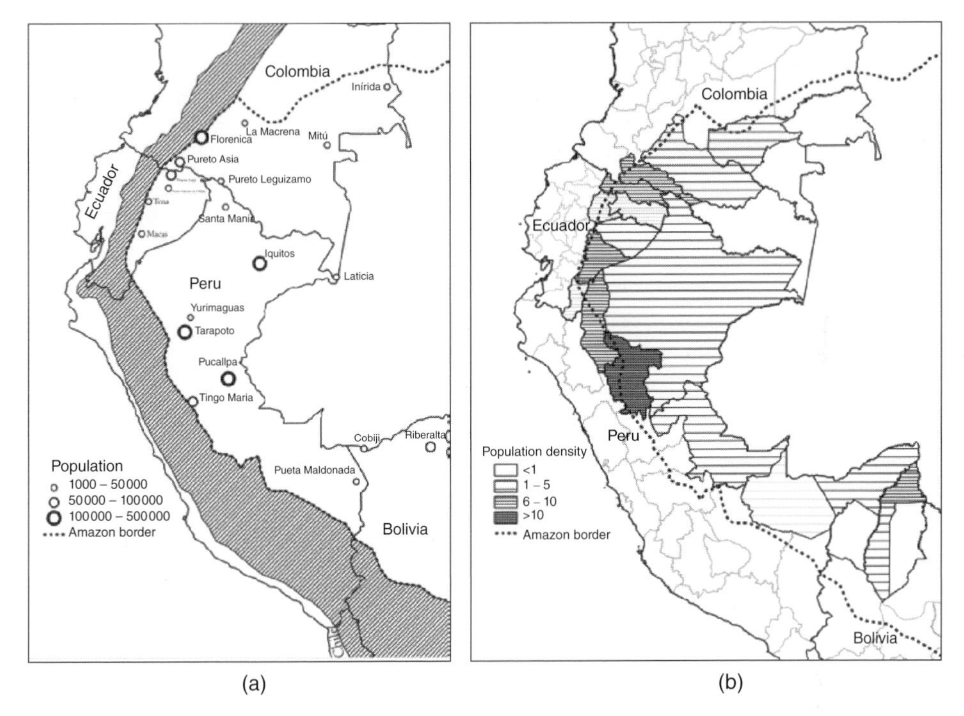

Figure 3.9 Population in Western Amazonia. (a) Major population centers (with >10 000 inhabitants) in the Amazonian departments/states/provinces (see Table 3.2). (b) Population density in the Amazonian departments.

in the Andean parts of all countries. In the lowlands, demographic changes may be more responsive to economic conjunctures than in the more populated parts of these countries. For example in the Peruvian lowland department of Loreto, which has recently suffered from economic difficulties due to lowered petroleum income, the natural population growth between the years 2010 and 2015 was 18,744 persons, but there was an average outmigration by 7,449 persons (INEI 2015). In the department of Madre de Dios, which is famous for its growth due to abundant natural resources and trade along the road connecting Amazonia with Andes, immigration statistics are positive, resulting into a yearly net population growth of 2.5%. In Ecuador there has also been a trend of decreasing population growth rates in the Amazonian provinces, yet growth rates are still high (during 2001 and 2010 they varied from 2.0% to 3.5% between different provinces).

Present Western Amazonian inhabitants represent a variety of different backgrounds and socio-economic potentials. Already the indigenous native groups are ethnically and linguistically heterogeneous since they descend from prehistoric peoples that have gone through territorial displacements, demographic changes and recent acculturation. Their traditional livelihood is closely related to the harvest of forest and riverine resources with subsistence farming and small-scale trading (Carneiro 1988). There are even some solitary groups living in voluntary isolation in the deep forests of Ecuador and in the borderlands

Table 3.2 Administrative units and population in Western Amazonia.

Country	Department/State/Province	Population	Capital
Colombia	Amazonas	46,950	Leticia
	Putumayo	237,197	Mocoa
	Caquetá	337,932	Cartagena de Chairá
	Guiania	18,797	Inirida
	Guaviare	56,758	San José del Guaviare
	Vaupés	19,943	Mitú
Ecuador	Orellana	136,396	Francisco de Orellana
	Pastaza	83,933	Puyo
	Napo	103,697	Tena
	Sucumbíos	176,472	Nueva Loja
	Morona Santiago	147,940	Macas
	Zamora Chinchipe	91,376	Zamora
Peru	Loreto	891,732	Iquitos
	San Martín	728,808	Moyobamba
	Ucayali	432,159	Pucallpa
	Amazonas	375,993	Chachapoyas
	Madre de Dios	109,555	Madre de Dios
Bolivia	Pando	128,944	Cobija
	Beni (prov. José Ballivián)	88,437	Trinidad[a)]
	Beni (prov. Vaca Diez)	122,140	
	Beni (prov Yacuma)	25,286	
	La Paz (prov. Abel Iturralde)	18,769	La Paz [a)]
TOTAL POPULATION		4,379 214	

a) not located in the Amazon region
Data sources (accessed on 25 November): Colombia (referred year 2005) http://www.dane.gov.co/index.php/esp/poblacion-y-registros vitales/censos/censo-2005; Ecuador (2010): http://www.ecuadorencifras.gob.ec/poblacion-y demografia/ Peru (2007): http://www.inei.gob.pe/estadisticas/indice-tematico/poblacion-y-vivienda/; Bolivia (2015): http://www.ine.gob.bo/indice/visualizador.aspx?ah=PC20103.HTM)

between Peru and Brazil. Although protected by special decrees, the survival of these peoples is threatened due to advancing oil production activities and illegal logging (Finer *et al.* 2008). A major group of rural settlers in Western Amazonia is however made up of the peoples called "ribereños", who mainly consist of mestizos that have inhabited the Amazonian floodplains for centuries. Their livelihood is based on the use of locally available resources and includes both subsistence economies and trade of agricultural and forest products (Kvist and Nebel 2001; Padoch 1988). The third group of rural inhabitants includes colonists from mid-1990s onwards. As they have mainly migrated to the lowlands from the Andes, the process is sometimes referred to as "Andeanisation of the Amazonia" (Pichón 1996). With respect, colonists usually lack traditional knowledge of the harvest, uses and management of Amazonia's natural resources. Their living is mainly based on small- and medium-scale farming or ranching on lands cleared of their original vegetation. There are also increasing numbers of pure urban dwellers who live in the midst of the rainforest area but have very thin contact with the surrounding wilderness. Cities with over 30,000 inhabitants in the region include Iquitos (ca. 370,000), Pucallpa (205,000), Tarapoto (145,000), Puerto Maldonado (74 000), Yurimaguas (63,000), Cobija (56 000), San José del Guaviare (45,000), Francisco de Orellana (45,000), Puyo (37,000), Tena (35,000) and Leticia (32,000). Their population constitutes some 30% of the total population in the region.

One could assume some sort of inter-regional integration within Western Amazonia, taken that the Andean–Amazonian interface is a natural uniting factor all across the region. Also the Spanish rule has left a strong legacy with shared administrative language and societal practices. Despite such strong similarities, regional transboundary integration is but faint. This situation is to a degree the result of many frontier disputes and conflicts, which have led to suspicious nationalistic relations among neighboring countries. Ironically, coca production and trade may be the most profitable and efficient human activity that has ever economically and culturally integrated countries in the Andean foothills and foreland belt (Franco and Godoy 1992; Whynes 1991).

All Western Amazonian countries have strong central administration and distinctively national communication infrastructures (Figure 3.10). Rivers provide the most natural transport network but as fluvial transportation takes a long time is not suitable for perishable goods, urgent deliveries or busy passengers (Salonen *et al.* 2012). The disposition of the fluvial system does not provide ideal support for modern regional integration either. Iquitos and Leticia are the only ports where ocean-going vessels can enter, both with lacking road connections to any other major population center. Many Western Amazonian rivers run to Brazil instead of forming unifying gateways within the region. Only in Peru, the Ucayali, Marañon and Amazonas rivers constitute a major fluvial network to join large Andean foreland areas together. This connection has undoubtedly helped regional integration in northern Peruvian Amazonia. A few navigable rivers also link Peru with Ecuador (the Pastaza and Napo rivers) but they have not resulted in notable transboundary benefits. Likewise, the Putumayo river is shared by Peru and Colombia, but it rather demarcates the national border than is used to gain bi-national synergy. In the south, the Madre de Dios river links southern Peru with northern Bolivia but also this connectivity potential is underused. Also the road networks in the region mainly support national needs. The only international road connections in the lowlands are between Peru and Brazil (from Puerto Maldonado to the state of Acre). Also the public airport network is sparse. Many airfields are only suitable for small plane traffic and not having regular flight connections. Regular commercial flights predominantly serve national transport needs within each country, whilst international connections between Western Amazonian cities are virtually lacking. Under-developed transport possibilities have without doubt contributed to the lack of regional integration, markets and identity in Western Amazonia.

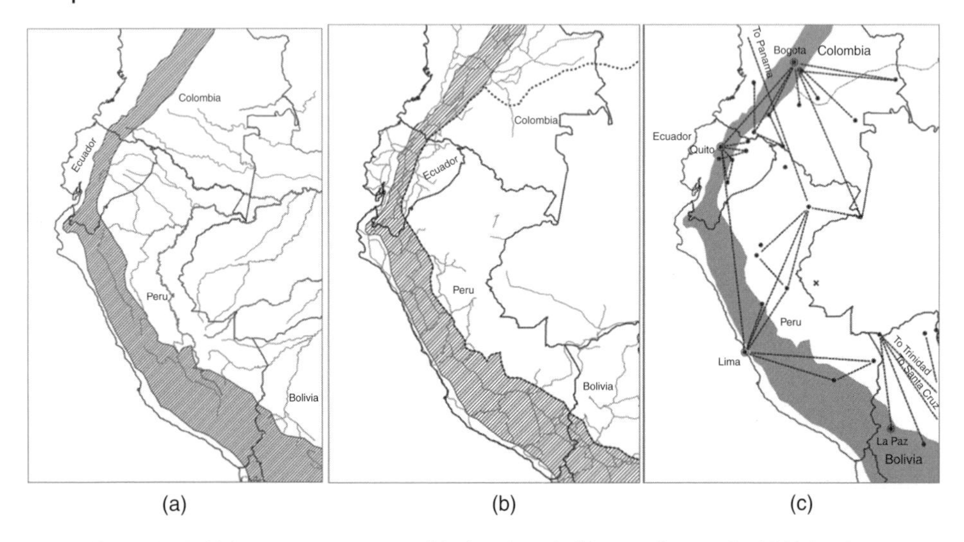

Figure 3.10 Major transport geographical settings in Western Amazonia. (a) Major river corridors providing fluvial interconnectivity. (b) Roads (source: http://raisg.socioambiental.org/mapa. (c) Airports and connections from Western Amazonian cities (source: http://www.flightconnections.com/).

Transboundary cooperation in Amazonia is forwarded by the Amazon Cooperation Treaty. Signed in 1978 and having a permanent secretariat in Brasilia, it offers a legal instrument to support harmonious development in the Amazon territories of Bolivia, Brazil, Colombia, Ecuador, Guyana, Peru, Suriname and Venezuela (ACTO 2015). Among its priorities is promotion for Ecological and Economic Macro-zoning, with special concern for the areas near national borders. Such zoning should lead to better integration of the Amazonian territories through methodologically comparable and mutually corresponding land use plans in different countries. The first efforts for ecological zoning had already started in the 1970s when Side-Looking Airborne Radar technology allowed imaging through clouds and was implemented in geomorphological, geological and vegetation inventories. This work was pioneered by the large-scale Brazilian government project RADAMBRASIL project (1973–1983) and later it was also applied to the western parts of Amazonia, for example by the Peruvian institute *Oficina Nacional de Evaluación de Recursos Naturales*. Also a number of other remote sensing technologies have been used to map the environment and resources in this region for promoting inventory-based zoning and planning. Having ecological, social and economic facts collected into a comprehensive sphere, stakeholders and expert groups are facilitated to create sound visions to guide future developments. Considering the complex environment in Western Amazonia, these requirements are, however, easier said than done. Inappropriate plans or their weak implementation may rather lead to overlooking of existing plans in case they are seen as too restrictive. Such a risk can be avoided by capacity building, broad scientific networking and institutional strengthening (Kainer *et al.* 2003; Kalliola *et al.* 2008). A solid knowledge base is particularly important since in contrast to most countries in the global North, planning in Amazonia is made on lands that still are in almost pristine condition.

The late 20th and early 21st centuries have also introduced new efforts to better integrate the Amazon Basin into the world economy. In particular, the Initiative for the Integration of the Regional Infrastructure of South America (IIRSA) is presented as a visionary program that aims to improve connectivity across South America by a paved inter-oceanic highway joining the two oceans through the Amazon. Although it is expected to stimulate economic growth in the region its likely consequences also include severe risk of large-scale environmental degradation, even in form of a drastic extinction event (Killeen 2007). In the tri-national frontier region of Bolivia, Brazil and Peru, increased road connectivity is already adding flows of people and goods and accelerates the likelihood of forest conversion by deforestation (Southworth *et al.* 2011). Improved infrastructures and stabilized democratic societies may consequently be opening resource use in Western Amazonia in an unprecedented way.

References

ACTO (2015) Amazon Cooperation Treaty Organization. http://otca.info/portal/index.php?p=index (accessed 18 November 2015).

Alonso, J.A. and Whitney, B.M. (2003) New distributional records of birds from white-sand forests of the northern Peruvian Amazon, with implications for biogeography of northern South America. *The Condor*, **105** (3), 552–566.

Arbeláez, M.V. and Duivenvoorden, J.F. (2004) Patterns of plant species composition on Amazonian sandstone outcrops in Colombia. *Journal of Vegetation Science*, **15** (2), 181–188.

Arnold, J.M. and Pérez, M.R. (2001) Can non-timber forest products match tropical forest conservation and development objectives? *Ecological Economics*, **39** (3), 437–447.

Asner, G.P., Llactayo, W., Tupayachi, R. and Luna, E.R. (2013) Elevated rates of gold mining in the Amazon revealed through high-resolution monitoring. *Proceedings of the National Academy of Sciences*, **110** (46), 18454–18459.

Bakewell, P. (1997) *A History of Latin America. Empires and Sequels 1450–1930*, Blackwell Publishers, Oxford, UK, p. 520.

Barham, B.L. and Coomes, O.T. (1996) *Prosperity's Promise – The Amazon Rubber Boom and distorted economic development*, Dellplain Latin American Studies Series, Westview Press, Boulder, CO, p. 179.

Becker, C.D. and León, R. (2000) Indigenous forest management in the Bolivian Amazon: lessons from the Yuracaré people, in *People and Forests: Communities, institutions, and governance* (eds C. Gibson, M. McKean and E. Ostrom), MIT Press, Cambridge, MA, pp. 163–191.

BIODAMAZ (2001) Estrategia regional del la diversidad biológica Amazónica, in *Documento técnico No. 01. Serie BIODAMAZ*, IIAP, Iquitos, p. 57.

Brown, L.A., Sierra, R., Digiacinto, S. and Smith, R. (1994) Urban-system evolution in frontier settings. *Geographical Review*, **84** (3), 249–266.

Burkholder, M.A. and Johnson, L.L. (1990) *Colonial Latin America*, Oxford University Press, New York, p. 400.

Bush, M.B., Silman, M.R. and Listopad, C.M.C.S. (2007) A regional study of Holocene climate change and human occupation in Peruvian Amazonia. *Journal of Biogeography*, **34** (8), 1342–1356.

Cardoso, E. and Fishbow, A. (1992) Latin American economic development: 1950–1980. *Journal of Latin American Studies. The colonial and post-colonial experience. Five centuries of Spanish and Portuguese America*, **24**(Quincentenary Supplement), 197–218.

Carneiro, R.L. (1988) Indians of the Amazonian forest, in *People of the Tropical Rain Forest* (eds J.S. Denslow and C. Padoch), University of California Press, London, pp. 73–87.

Carneiro da Cunha, M. and de Almeida, M.W.B. (2000) Indigenous people, traditional people and conservation in the Amazon. *Daedalus*, **129** (2), 315–338.

Cleary, D. (2001) Towards an environmental history of the Amazon: from prehistory to the nineteenth century. *Latin American Research Review*, **36** (2), 65–96.

Coomes, O. (1995) A century of rain forest use in Western Amazonia. Lessons for extraction-based conservation of tropical forest resources. *Forest and Conservation History*, **39** (3), 108–120.

Coomes, O. (1996) State credit programs and the peasantry under populist regimes: Lessons from the APRA experience in the Peruvian Amazon. *World Development*, **24** (8), 1333–1346.

DANE (Departamento Administrativo Nacional de Estadística) (1993) XVI Censo Nacional de Población y V de Vivienda. http://www.dane.gov.co/ (accessed 30 January 2001).

De Boer, W.R. (1981) The machete and the cross: Conibo trade in the late seventeeth century, in *Networks of the Past: Regional interaction in archaeology* (eds P.D. Francis, F.J. Kense and P.G. Dahe), Archaeological Association, University of Galgary, Francis.

Denevan, W.M. (1996) A bluff model of riverine settlement in prehistoric Amazonia. *Annals of Association of American Geographers*, **86** (4), 654–681.

Denevan, W.M. (2001) *Cultivated Landscapes of Native Amazonia and the Andes*, Oxford University Press, New York, p. 396.

Eastwood, D.A. and Pollard, H.J. (1992) Amazonian colonization in Eastern Ecuador: Land use conflicts in a planning vacuum. *Singapore Journal of Tropical Geography*, **13** (2), 103–117.

Ebeling, J. and Yasué, M. (2009) The effectiveness of market-based conservation in the tropics: Forest certification in Ecuador and Bolivia. *Journal of Environmental Management*, **90** (2), 1145–1153.

Erickson, C.L. (1995) Archaeological perspectives on ancient landscapes of the Llanos de Mojos in the Bolivian Amazon, in *Archaeology in the American Tropics: Current Analytical Methods and Applications* (ed. Stahl), Cambridge University Press, Cambridge, UK, pp. 66–95.

Erickson, C.L. (2000) An artificial landscape-scale fishery in the Bolivian Amazon. *Nature*, **408**, 190–193.

Erickson, C.L. (2010) The transformation of environment into landscape: the historical ecology of monumental earthwork construction in the Bolivian Amazon. *Diversity*, **2** (4), 618–652.

Fagan, B.M. (1987) *The Great Journey: The Peopling of Ancient America*, Thames and Hudson Ltd, London, p. 112.

Finer, M. and Orta-Martínez, M. (2010) A second hydrocarbon boom threatens the Peruvian Amazon: trends, projections, and policy implications. *Environmental Research Letters*, **5** (1), 014012.

Finer, M., Jenkins, C.N., Pimm, S.L. *et al.* (2008) Oil and gas projects in the western Amazon: threats to wilderness, biodiversity, and indigenous peoples. *PLoS ONE*, **8**, e2932.

Foley, J.A., Asner, G.P., Costa, M.H. *et al.* (2007) Amazonia revealed: forest degradation and loss of ecosystem goods and services in the Amazon Basin. *Frontiers in Ecology and the Environment*, **5** (1), 25–32.

Franco, M. and Godoy, R. (1992) The economic consequences of cocaine production in Bolivia: historical, local and macroeconomic perspectives. *Journal of Latin American Studies*, **24**, 375–406.

French-Davis, R., Muñoz, O. and Palma, J.G. (1998) The Latin American economies, 1950–1990, in *Latin America: Economy and society since 1930* (ed. L. Bethell), Cambridge University Press, Cambridge, UK, pp. 149–237.

Garreaud, R.D. (2009) The Andes climate and weather. *Advances in Geosciences*, **22** (22), 3–11.

Gentry, A.H., Churchill, S.P., Balslev, H. *et al.* (1995) Patterns of diversity and floristic composition in Neotropical montane forests, in *Biodiversity and Conservation of Neotropical Montane Forests. Proceedings of Symposium, New York Botanical Garden, 21–26 June 1993*, New York Botanical Garden, New York, pp. 103–126.

Gibbons, A. (1990) New view of early Amazonia. *Science*, **248**, 1488–1490.

Giudice, R., Soares-Filho, B.S., Merry, F. *et al.* (2012) Timber concessions in Madre de Dios: Are they a good deal? *Ecological Economics*, **77**, 158–165.

Gómez, N.E. (1996) *Elementos de geografía del Ecuador*, El hombre y el medio. Ediguias C. Ltda, Quito, p. 190.

Haring, C.H. (1973) *The Spanish Empire in America*, Harcourt Brace Jovanovich, Inc, San Diege, CA.

Higbee, E.C. (1951) Of man and the Amazon. *The Geographical Review*, **41** (3), 410–420.

Higgins, M.A., Ruokolainen, K., Tuomisto, H. *et al.* (2011) Geological control of floristic composition in Amazonian forests. *Journal of Biogeography*, **38** (11), 2136–2149.

Hoorn, C. (1994) An environmental reconstruction of the palaeo-Amazon river system (Middle–Late Miocene, NW Amazonia). *Palaeogeography, Palaeoclimatology, Palaeoecology*, **112** (3), 187–238.

Hoorn, C., Wesselingh, F.P., Hovikoski, J. and Guerrero, J. (2010) The development of the Amazonian mega-wetland (Miocene; Brazil, Colombia, Peru, Bolivia), in *Amazonia, Landscape and Species Evolution: A Look into the past*, Blackwell-Wiley, Hoboken, pp. 123–142.

INEI (2015) Indicadores demográficos, por departamento, 2010–2015. http://www.inei.gob.pe (accessed 17 November 2015).

Irion, G. and Kalliola, R. (2009) Long-term landscape development processes in Amazonia, in *Amazonia, Landscape and Species Evolution. A look into the past* (eds C. Hoorn and F. Wesselingh), Wiley-Blackwell, Oxford, UK, pp. 185–197.

ITTO (2009) Encouraging industrial forest plantations in the tropics. Report of a global study. *ITTO Technical Series*, **33**, 1–141.

Jorge Pádua, M.T., Magnanini, A. and Mittermeier, R.A. (1974) Brazils national parks. *Oryx*, **12** (2), 452–461.

Kainer, K.A., Schmink, M., Pinheiro Leite, A.C. and da Silva Fadell, M.J. (2003) Experiments in forest-based development in western Amazonia. *Society and Natural Resources*, **16** (10), 869–886.

Kalliola, R. (2011) Brazil nut harvesting in Peruvian Amazonia from the perspective of ecosystem services. *Fennia*, **189** (2), 1–13.

Kalliola, R., Puhakka, M., Salo, J. *et al.* (1991) The dynamics, distribution and classification of swamp vegetation in Peruvian Amazonia. *Annales Botanici Fennici*, **28**, 225–239.

Kalliola, R., Salo, J., Puhakka, M. *et al.* (1992) Upper Amazon channel migration: implications for vegetation perturbance and succession using bitemporal Landsat MSS images. *Naturwissenschaften*, **79**, 75–79.

Kalliola, R., Toivonen, T., Miyakawa, V. and Mavila, M. (2008) Open access to information bridges science and development in Amazonia: lessons of the SIAMAZONIA service. *Environmental Research Letters*, **3**, 034004.

Karsten, R.J., Meilby, H. and Larsen, J.B. (2014) Regeneration and management of lesser known timber species in the Peruvian Amazon following disturbance by logging. *Forest Ecology and Management*, **327**, 76–85.

Khazeh, K. and Clark, D.B. (1990) A case study of effects of developing country integration on trade flows: the Andean Pact. *Journal of Latin American Studies*, **22**, 317–330.

Killeen, T.J. (2007) A perfect storm in the Amazon wilderness. *Advances in Applied Biodiversity Science*, **7**, 102.

Kirkby, C.A., Giudice, R., Day, B. *et al.* (2011) Closing the ecotourism-conservation loop in the Peruvian Amazon. *Environmental Conservation*, **38** (01), 6–17.

Kvist, L.P. and Nebel, G. (2001) A review of Peruvian flood plain forests: ecosystems, inhabitants and resource use. *Forest Ecology and Management*, **150**, 3–26.

Lähteenoja, O., Ruokolainen, K., Schulman, L. and Alvarez, J. (2009) Amazonian floodplains harbour minerotrophic and ombrotrophic peatlands. *Catena*, **79** (2), 140–145.

Larrea, C. and North, L.L. (1997) Ecuador: Adjustment policy impacts on truncated development and democratisation. *Third World Quarterly*, **18** (5), 913–934.

Lathrap, D.W. (1973) The antiquity and importance of long-distance trade relationships. *World Archaeology*, **5** (2), 170–186.

Lombardi, C.L., Lombardi, J.V. and Stoner, K.L. (1983) *Latin American History. A teaching atlas*, The University of Wisconsin Press, London.

Lynch, J. (1992) The institutional framework of colonial Spanish America. The colonial and post-colonial experience. Five centuries of Spanish and Portuguese

America. *Journal of Latin American Studies*, **24**(Quincentenary Supplement), 69–81.

Mäki, S., Kalliola, R. and Vuorinen, K. (2001) Road construction in the Peruvian Amazon: process, causes and consequences. *Environmental Conservation*, **28** (03), 199–214.

Marengo, J.A. (1992) Interannual variability of surface climate in the Amazon Basin. *International Journal of Climatology*, **12** (8), 853–863.

McMichael, C.H., Palace, M.W., Bush, M.B. *et al.* (2014) Predicting pre-Columbian anthropogenic soils in Amazonia. *Proceedings of the Royal Society of London B: Biological Sciences*, **281** (1777), 2013–2475.

Meertens, D. and Segura-Escobar, N. (1996) Uprooted lives: gender, violence and displacement in Colombia. *Singapore Journal of Tropical Geography*, **17** (2), 165–178.

Mejia, E., Pacheco, P., Muzo, A. and Torres, B. (2015) Smallholders and timber extraction in the Ecuadorian Amazon: amidst market opportunities and regulatory constraints. *International Forestry Review*, **17** (S1), 38–50.

Messina, J.P., Walsh, S.J., Mena, C.F. and Delamater, P.L. (2006) Land tenure and deforestation patterns in the Ecuadorian Amazon: conflicts in land conservation in frontier settings. *Applied Geography*, **26** (2), 113–128.

Ministerio de Relaciones Exteriores del Perú (2002) Plan binacional de desarollo de la región fronteriza Perú – Ecuador. http://www.rree.gob.pe/ (accessed 14 December 2002).

Murphy, L., Bilsborrow, R. and Pichón, F. (1997) Poverty and prosperity among migrant settlers in the Amazon rainforest frontier of Ecuador. *The Journal of Development Studies*, **34** (2), 35–66.

Myster, R.W. (2009) Plant communities of Western Amazonia. *The Botanical Review*, **75** (3), 271–291.

Nebel, G., Kvist, L.P., Vanclay, J.K. and Vidaurre, H. (2001) Forest dynamics in flood plain forests in the Peruvian Amazon: effects of disturbance and implications for management. *Forest ecology and Management*, **150** (1), 79–92.

Nelson, B.W. (1994) Natural forest disturbance and change in the Brazilian Amazon. *Remote Sensing Reviews*, **10** (1–3), 105–125.

Nepstad, D.C., Verssimo, A., Alencar, A. *et al.* (1999) Large-scale impoverishment of Amazonian forests by logging and fire. *Nature*, **398** (6727), 505–508.

Oliveira, P.J., Asner, G.P., Knapp, D.E. *et al.* (2007) Land-use allocation protects the Peruvian Amazon. *Science*, **317** (5842), 1233–1236.

Padoch, C. (1988) People of the floodplain and forest, in *People of the Tropical Rain Forest* (eds J.S. Denslow and C. Padoch), University of California Press, London, pp. 127–140.

Pärssinen, M. (1992) *Tawantisuyu. The Inca State and its political organisation*, Gummerus Oy, Jyväskylä, p. 462.

Pärssinen, M., Schaan, D. and Ranzi, A. (2009) Pre-Columbian geometric earthworks in the upper Purús: a complex society in Western Amazonia. *Antiquity*, **83** (322), 1084–1095.

Peres, C.A. and Terborgh, J.V. (1995) Amazonian nature reserves: an analysis of the defensibility status of existing conservation units and design criteria for the future. *Conservation biology,* **9** (1), 34–46.

Philip, G. (2001) Commodities in Latin America, in *South America, Central America and the Caribbean 2002, Regional Surveys of the World* (ed. J. West), Europa Publications, Taylor & Francis Group, London, pp. 26–29.

Pichón, F.J. (1996) Land-use strategies in the Amazon frontier: farm level evidence from Ecuador. *Human Organization,* **55** (4), 416–424.

Pinedo-Vasquez, M., Zarin, D.J., Coffey, K. *et al.* (2001) Post-boom logging in Amazonia. *Human Ecology,* **29** (2), 219–239.

Pokorny, B., Scholz, I. and de Jong, W. (2013) REDD+ for the poor or the poor for REDD+? About the limitations of environmental policies in the Amazon and the potential of achieving environmental goals through pro-poor policies. *Ecology and Society,* **18** (2), 3.

Puhakka, M., Kalliola, R., Rajasilta, M. and Salo, J. (1992) River types, site evolution and successional vegetation patterns in Peruvian Amazonia. *Journal of Biogeography,* **19**, 651–665.

RAISG (2012) Amazonía bajo preción. 68 págs. (www.raisgsocioambiental.org)

Räsänen, M.E., Salo, J.S. and Kalliola, R.J. (1987) Fluvial perturbance in the Western Amazon Basin: regulation by long-term sub-Andean tectonics. *Science,* **238**, 1398–1401.

Räsänen, M., Neller, R., Salo, J. and Jungner, H. (1992) Recent and ancient fluvial deposition systems in the Amazonian foreland basin, Peru. *Geological Magazine,* **129** (3), 293–306.

Räsänen, M.E., Linna, A.M., Santos, J.C.R. and Negri, F.R. (1995) Late Miocene tidal deposits in the Amazonian foreland basin. *Science,* **269**, 386–390.

Reeve, M.-E. (1994) Regional interaction in the Western Amazon: the early colonial encounter and the Jesuit years: 1538–1767. *Ethnohistory,* **41** (1), 106–138.

Roosevelt, A. (1996) Paleoindian cave dwellers in the Amazon: the peopling of the Americas. *Science,* **272**, 373–384.

Rudel, T.K. (1983) Roads, speculators and colonization in the Ecuadorian Amazon. *Human Ecology,* **11** (4), 385–403.

Salo, J., Kalliola, R., Häkkinen, I. *et al.* (1986) River dynamics and the diversity of Amazon lowland forest. *Nature,* **322**, 254–258.

Salo, M. and Toivonen, T. (2009) Tropical timber rush in Peruvian Amazonia: spatial allocation of forest concessions in an uninventoried frontier. *Environmental Management,* **44** (4), 609–623.

Salo, M., Helle, S. and Toivonen, T. (2011) Allocating logging rights in Peruvian Amazonia – does it matter to be local. *PloS one,* **6** (5), e19704.

Salo, M., Sirén, A. and Kalliola, R. (2014) *Diagnosing Wild Species Harvest Resource Use and Conservation,* Academic Press (Elsevier), Oxford, UK, p. 479.

Salonen, M., Toivonen, T., Cohalan, J.M. and Coomes, O.T. (2012) Critical distances: comparing measures of spatial accessibility in the riverine landscapes of Peruvian Amazonia. *Applied Geography,* **32** (2), 501–513.

San Sebastián, M. and Hurtig, A.K. (2005) Oil development and health in the Amazon basin of Ecuador: the popular epidemiology process. *Social Science and Medicine,* **60** (4), 799–807.

Schulman, L., Ruokolainen, K., Junikka, L. *et al.* (2007) Amazonian biodiversity and protected areas: do they meet? *Biodiversity and Conservation*, **16** (11), 3011–3051.

Schuurman, F.J. (1979) Colonization policy and peasant economy in the Amazon basin. *Boletin de Estudios Latinoamericanos y del Caribe*, **27**, 29–41.

Sempat Assadourian, C. (1992) The colonial economy: The transfer of the European system of production to New Spain and Peru. *Journal of Latin American Studies. The colonial and post-colonial experience. Five centuries of Spanish and Portuguese America*, **24** (Quincentenary Supplement), 55–68.

Shearman, P., Bryan, J. and Laurance, W.F. (2012) Are we approaching "peak timber" in the tropics? *Biological Conservation*, **151** (1), 17–21.

Skole, D. and Tucker, C. (1993) Tropical deforestation and habitat fragmentation in the Amazon. Satellite data from 1978 to 1988. *Science*, **260** (5116), 1905–1910.

Smith, A. (1990) *Explorers of the Amazon*, The University of Chicago Press, Chicago, IL, p. 344.

Southworth, J., Marsik, M., Qiu, Y. *et al.* (2011) Roads as Drivers of Change: trajectories across the Tri-National Frontier in MAP, the Southwestern Amazon. *Remote Sensing*, **3** (5), 1047–1066.

Steinberg, M.K. (2000) Generals, guerillas, drugs, and Third World war-making. *Geographical Review*, **90** (2), 260–268.

Sun, J., Southworth, J. and Qiu, Y. (2015) Mapping multi-scale impacts of deforestation in the Amazonian rainforest from 1986 to 2010. *Journal of Land Use Science*, **10** (2), 174–190.

Sustainable Forest Products (2015) http://www.sustainableforestprods.org/home (accessed 23 October 2015).

Thiele, G. (1995) The displacement of peasant settlers in the Amazon: the case of Santa Cruz, Bolivia. *Human Organization*, **54** (3), 273–282.

Toivonen, T., Mäki, S. and Kalliola, R. (2007) The riverscape of Western Amazonia – a quantitative approach to the fluvial biogeography of the region. *Journal of Biogeography*, **34** (8), 1374–1387.

Tuomisto, H. (2007) Interpreting the biogeography of South America. *Journal of Biogeography*, **34** (8), 1294–1295.

Tuomisto, H., Ruokolainen, K., Kalliola, R. *et al.* (1995) Dissecting Amazonian biodiversity. *Science*, **269**, 63–66.

Vasquez, R. and Gentry, A.H. (1989) Use and misuse of forest-harvested fruits in the Iquitos Area. *Conservation biology*, **3** (4), 350–361.

Vormisto, J., Svenning, J.C., Hall, P. and Balslev, H. (2004) Diversity and dominance in palm (Arecaceae) communities in *terra firme* forests in the western Amazon basin. *Journal of Ecology*, **92** (4), 577–588.

Walker, R. and Smith, T.E. (1993) Tropical deforestation and forest management under the system of concession logging: a decision-theoretic analysis. *Journal of Regional Science*, **33** (3), 387–419.

West, J. (2001) *South America, Central America and the Caribbean 2002*, Regional Surveys of the World. Europa Publications. Taylor & Francis Group, London, p. 930.

Whynes, D. (1991) Illicit drug production and supply-side drugs policy in Asia and South America. *Development and Change*, **22**, 475–496.

Wilgus, A.C. (1967) *Historical Atlas of Latin America*, Cooper Square Publishers, New York, p. 365.

WMO/UNESCO (1975) *Climatic Atlas of South America* Maps of mean temperature and precipitation, Cartographia, Budapest.

World Resources Institute (2015) Zooming In: "Sustainable" cocoa producer destroys pristine forest in Peru. http://www.wri.org/blog/2015/06/zooming-%E2%80%9Csustainable%E2%80%9D-cocoa-producer-destroys-pristine-forest-peru (accessed 25 November 2015).

Zeidemann, V., Kainer, K.A. and Staudhammer, C.L. (2014) Heterogeneity in NTFP quality, access and management shape benefit distribution in an Amazonian extractive reserve. *Environmental Conservation*, **41** (03), 242–252.

4

Forest Structure, Fruit Production and Frugivore Communities in *Terra firme* and *Várzea* Forests of the Médio Juruá

Joseph E. Hawes and Carlos A. Peres

Abstract

Floodplain forests comprise some of the defining and most enigmatic habitats of Western Amazonia. This chapter provides an overview of the multi-directional relationships between: forest structure and carbon stocks; fruit production and phenology; and frugivores and seed dispersal services, paying particular attention throughout to the role of the flood pulse by examining areas of adjacent *varzea* and *terra firme* forest. Examining the impact of the seasonal flood on plant communities and phenological patterns in fruit production, and the relative importance of different seed dispersal modes, helps interpret the respective frugivore communities of *terra firme* and *varzea* forests. Contrasting the differences found between these adjacent forest types adds an important perspective to the individual assessments of above-ground biomass, phenological patterns and fruit-frugivore interactions, and these multiple components also combine in speculation over the possible implications of an 'empty flooded forest'. This research helps to address the shortage of such studies in Western Amazonia to date.

Keywords *carbon stocks; empty flooded forest; floodplain forests; forest structure; frugivores; fruit production; seed dispersal services; terra firme forests; varzea forests; Western Amazonia*

4.1 Introduction

Floodplain forests comprise some of the defining and most enigmatic habitats of Western Amazonia. Of the seven major wetland types identified across Amazonia (Pires and Prance 1985), *várzea* forests are the most extensive, accounting for more than 200,000 km^2 within Brazilian Amazonia alone (Junk 1997) and more than 400,000 km^2 in total (Junk *et al.* 2011; Melack and Hess 2010). They are defined as the white-water floodplains of the Amazon (=Solimões) river and its tributaries (Prance 1979), and can be inundated for up to 230 days per year, at depths rising to 7.5 m (Junk *et al.* 2011). The "white-water" of these rivers is derived from their high load of Andean alluvial sediments (Irion *et al.* 1997), of which 300–1,000 mm of erosional nutrient-rich deposits can be contributed

Forest structure, function and dynamics in Western Amazonia, First Edition.
Edited by Randall W. Myster.

to *várzea* soils every year (Parolin 2009). Seasonal floodwaters result in high fertility (Sioli 1984) and primary/secondary productivity levels two to three times higher than in adjacent heavily leached and nutrient poor *terra firme* forests (Worbes 1997).

In addition to the high fertility of *várzea* forests, the regular annual "flood pulse" (Junk 1989) has many additional consequences. The extended period of submersion and waterlogging alternates with contrasting drought conditions when the floods retreat (Parolin *et al.* 2010a), resulting in clearly demarcated terrestrial and aquatic phases. This cycle plays a fundamental role as a selective pressure on a range of structural, physiological and phenological adaptations within the plant community (Parolin *et al.* 2004b) and can help determine community composition and explain many life-history traits of *várzea* tree species, including growth rates, wood density, phenological strategies and fruit/seed morphology.

Forest structure and composition combine to determine levels of above-ground biomass (AGB), a property of forests with increasing relevance in a world concerned with global carbon stocks. The physiological adaptations within individual species and resulting life-history traits, driven by the extreme environmental conditions, are also strongly influential on AGB estimates. For example, the hyper-abundant nutrient conditions in the disturbance-prone *várzea* environment favors fast life histories of short-lived individuals with rapid growth rates, frequently resulting in low wood densities (Baker *et al.* 2004; Fearnside 1997). Coupled with structural differences in stem density, basal area and tree height, this suggests that AGB estimates produced from *terra firme* forest plots cannot be reliably extrapolated across *várzea* forest.

In addition to carbon stocks, forest structure and composition have a direct influence on resource production, including the timber and non-timber products exploited by humans for commercial and subsistence purposes. Plants also provide an important resource for a wide range of vertebrate taxa (Fleming and Kress 2011; Smythe 1986), with frugivores particularly ubiquitous in tropical forests. Fruit–frugivore interactions represent a mutually beneficial relationship between vertebrates and plants, which has developed through a long co-evolutionary process over 90 Ma (Fleming and Kress 2011). In *várzea* forests, frugivores have needed to adapt to variation in the availability of fruits both spatially, within and between adjacent *terra firme* and *várzea* forests, and temporally, in relation to annual fluctuations in environmental factors including rainfall and the flood pulse. Frugivores are also likely to be influenced by plant adaptations in response to the seasonal inundation, including seed dispersal modes that take advantage of the prolonged flood pulse. Animal-dispersed plants bearing fleshy fruits are well represented in humid tropical forests in general, and particularly so in Western Amazonian forests, while abiotically dispersed plants are typically more common in dry forests (Griz and Machado 2001). In seasonally-inundated forests such as *várzea*, a higher proportion of plants bear seeds dispersed by abiotic agents, in particular water, which is expected to be one of the main dispersal vectors (Kubitzki and Ziburski 1994; Oliveira and Piedade 2002; Parolin *et al.* 2010b, 2013). However, this is likely to be tempered to some extent by the importance of fish in seed dispersal in flooded forests (Horn *et al.* 2011).

Seed dispersal is a crucial component of a functioning ecosystem (Levin *et al.* 2003; Nathan and Muller-Landau 2000) and the loss of links from these mutualistic networks may have potentially catastrophic cascading effects (Wright 2003; Wright *et al.* 2007). There is now considerable attention focused on the resilience of tropical forests to cope with the loss of large-bodied frugivores (Peres 2000; Terborgh *et al.* 2008), with their local depletion and extirpation turning the once envisioned "empty forest" scenario into a reality (Redford 1992; Wilkie *et al.* 2011). Large-bodied frugivores are at greater risk from selective hunting, which could threaten the demographic viability of large-fruited or large-seeded plants (Peres and van Roosmalen 2002; Wheelwright 1985), unless alternative frugivores can effectively provide substitutional roles as dispersal agents. Determining the variation in fruit trait selection and degree of dietary overlap in co-existing consumers is therefore critical to understanding potential frugivore resilience to disturbance. This importance is further enhanced by the high wood densities typical of many large-seeded plant species, which mean that any reduced recruitment as a result of missing seed dispersal agents could have serious implications for above-ground biomass and carbon stocks in tropical forests (Bello *et al.* 2015 Levin *et al.* 2003; Peres *et al.* 2016).

This chapter aims to provide an overview of the multi-directional relationships between:

1. forest structure and carbon stocks;
2. fruit production and phenology; and
3. frugivores and seed dispersal services,
   paying particular attention throughout to the role of the flood pulse by examining areas of adjacent *várzea* and *terra firme* forest.

Examining the impact of the seasonal flood on plant communities and phenological patterns in fruit production, and the relative importance of different seed dispersal modes, helps interpret the respective frugivore communities of *terra firme* and *várzea* forests. Contrasting the differences found between these adjacent forest types adds an important perspective to the individual assessments of above-ground biomass, phenological patterns and fruit–frugivore interactions, and these multiple components also combine in speculation over the possible implications of an "empty flooded forest".

Finally, this research helps to address the shortage of such studies in Western Amazonia to date. For example, the distribution of *várzea* forest inventories is patchy and concentrated around major urban centers (Albernaz *et al.* 2012), as for Amazonian forests in general (Hopkins 2007), and the small areas of *várzea* sampled to date throughout Amazonia are unlikely to be representative, with vast regions remaining entirely unknown (Parolin *et al.* 2004a). In particular, few *várzea* studies have been conducted between existing plot-scale inventories in central Brazilian Amazonia (e.g. Assis and Wittmann 2011; Ayres 1986; Schöngart *et al.* 2010; Worbes 1997) and those in the upper Ecuadorian, Bolivian and Peruvian Amazon (e.g. Balslev *et al.* 1987; Comiskey *et al.* 2000; Freitas Alvarado 1996; Gentry 1988; Myster 2007, 2014; Nebel *et al.* 2001). The only studies we are aware of within this major gap in lowland Western Amazonia are those in Rodrigues Alves, Acre (Campbell *et al.* 1992) and the Médio Juruá,

Amazonas (Assis *et al.* 2014, 2015; Hawes *et al.* 2012; Peres and Malcom, unpublished data). Few quantitative assessments of community-wide plant phenology are available either (Parolin *et al.* 2010c), with most studies again located in close proximity to Manaus and focusing on a select few tree species (Hawes and Peres 2016). With the exception of one other study (Haugaasen and Peres 2005), which also compares the phenology patterns of *várzea* to *igapó* (black-water flooded forests), there exists a distinct lack of direct comparisons between *várzea* and *terra firme* forests.

Similarly, although several tropical forest studies have examined differences in fruit trait selection within a single frugivore assemblage (Bollen *et al.* 2004; Donatti *et al.* 2011; Flörchinger *et al.* 2010; Gautier-Hion *et al.* 1985; Kitamura *et al.* 2002; Voigt *et al.* 2004), surprisingly few have been attempted in lowland Amazonia (Link and Stevenson 2004), even though this region holds both the highest diversity of terrestrial and aquatic frugivorous vertebrates (Fleming *et al.* 1987) and the widest spectrum of morphological fruit types (Gentry 1996; van Roosmalen 1985) anywhere on Earth. This is particularly the case for seasonally flooded *várzea* forests, which are the most species-rich floodplain forests worldwide (Wittmann *et al.* 2006) and, although comprising only approximately 5% of Amazonia, account for as much as 20% of its woody flora (Junk 1997).

4.2 Methods

Our study area in the western Brazilian Amazon is located within two contiguous sustainable-use forest reserves in the state of Amazonas, encompassing nearly 0.9 Mha: the Médio Juruá Extractive Reserve (*ResEx Médio Juruá*, 253,227 ha) and the Uacari Sustainable Development Reserve (*RDS Uacari*, 632,949 ha) (Figure 4.1). These two reserves border the Juruá river, a major white-water tributary of the Solimões (=Amazon) river, and contain large expanses of upland unflooded *terra firme* forest (80.6% of combined reserve area) and seasonally-flooded *várzea* forest (17.9%) closer to the main river channel (Hawes *et al.* 2012). *Terra firme* soils are typically heavily leached and nutrient poor in comparison to the eutrophic soils of *várzea* forests (Furch 1997), which are renewed with a fresh layer of pre-Andean alluvial sediments every year. The elevation range within the study area is 65–170 m a.s.l. and all sites surveyed consisted of undisturbed primary forest.

The Médio Juruá region has a wet, tropical climate with marked seasonal variation in rainfall, temperature, humidity and floodwaters. There is a mean annual temperature of 27.1°C and annual rainfall averages 3,679 mm/yr, based on daily records over three consecutive years (2008–2010) at the Bauana Ecological Field Station (5°26′19″S, 67°17′12″W) and the Eirunepé meteorological station (315 km from the study area, 2000–2010, source: INMET). Although hot and humid throughout the year, the hottest months are August–November, and humidity peaks in January–April. The precipitation pattern (rainy season: November–April, dry season: May–October) is asynchronous with the flood

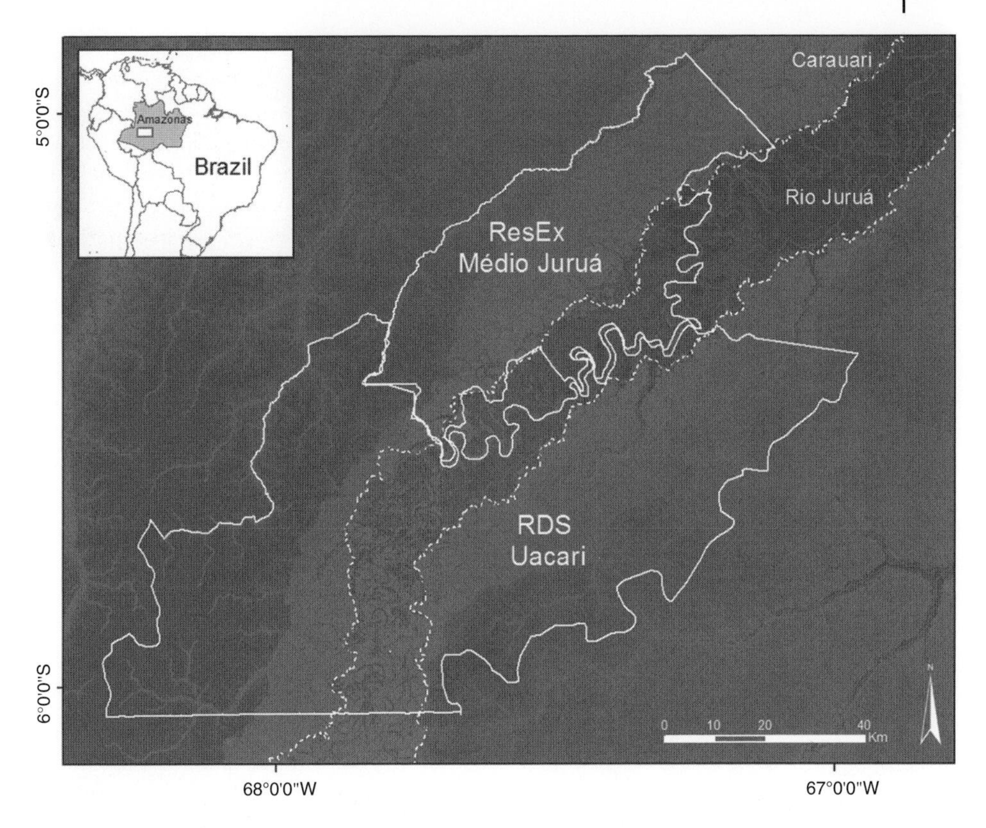

Figure 4.1 Map of the Médio Juruá region of western Brazilian Amazonia, showing the distribution of forest types within the two study reserves. Colors indicate terrain elevation, which corresponds approximately with the boundary between *terra firme* and *várzea* forests more clearly shown by the dashed lines.

pulse, so that the river and floodplain water-levels lag approximately 6 weeks behind rainfall (aquatic phase: January–June, terrestrial phase: July–December) (Hawes and Peres 2016).

To assess forest structure and above-ground biomass, 20 ha of forest in the Médio Juruá were sampled using 200 tree plots (0.1-ha, 100 m × 10 m), with two sets of 100 plots divided equally across *terra firme* and *várzea* forests and survey effort also divided equally between the left and right banks of the Juruá river (see Hawes *et al.* 2012 for further details). Within each of our plots, all live stems (including palms but excluding woody lianas and non-freestanding hemi-epiphytes) ≥10 cm in diameter at breast height (dbh) were identified and measured. Vernacular names were attributed to the highest possible level of taxonomic resolution (species 18.4%, genus 59.8%, family 19.5%), with only 2.4% of all trees (N = 12,721) within the 200 plots remaining unidentified.

To monitor the vegetative and reproductive plant phenology of *várzea* floodplain and adjacent *terra firme* forests, three complementary methods were used: monthly canopy observations of 1,056 individuals (TF: 556, VZ: 500), twice monthly collections from 0.5 m^2 litterfall traps within two 100-ha plots (1 TF,

Figure 4.2 Comparative views of *terra firme* and várzea forests in the Médio Juruá region of western Brazilian Amazonia, and corresponding field methods in each forest type.

1 VZ; 96 traps per plot), and monthly ground surveys of residual fruit-fall along transect grids within each 100-ha plot (12 km per plot). Surveys encompassed the entire annual flood cycle and employed a floating trap design to cope with fluctuating water levels (Figure 4.2, see Hawes and Peres 2016 for further details).

Finally, a synthesis of field and interview methods were employed to compare the plant diets of medium- to large-bodied terrestrial, arboreal and aquatic frugivorous vertebrates in both *várzea* and *terra firme* forest, and examine the relative contribution of dispersal modes and other fruit traits, including fruit morphology and color, to their diet selection in terms of fruit resources. Monthly surveys of fruit patches and medium- to large-bodied vertebrate frugivores were conducted within three 100-ha plots (2 TF, 1 VZ), supplemented by fruit surveys along 67 transects distributed across the study forest reserves (41 TF, 26 VZ; 5 km per transect). Observations of trophic interactions were augmented by semi-structured interviews with experienced hunters and fishermen from 16 local communities. To our knowledge, this represents the first systematic attempt to comprehensively document the tropical fruit–frugivore networks of two adjacent, yet radically different, forest types (see Hawes and Peres 2014 for further details).

4.3 Results and discussion

Stem density in the Médio Juruá was similar in *terra firme* and *várzea* forests, with both forest types dominated by smaller stems, although large emergent trees (>100 cm dbh) had a disproportionately large influence on plot basal area, particularly in *várzea* forests. As a result, mean plot basal area was greater and more variable in *várzea* than in *terra firme* forest (Hawes *et al.* 2012). This finding is similar to that in the Ecuadorian Amazon, where it is suggested to be driven by a greater flood-induced mortality in smaller stems (Myster 2014). Flooding also contributes to other structural characteristics of *várzea* forests, including canopy height and openness. These properties have not yet been explored in the *várzea* forests of the Médio Juruá, but it is recognized that unstable soils coupled with the persistent flood pulse promote a high incidence of tree-fall gaps, reducing competition for light, and substantially lowering the canopy stature in comparison to *terra firme* forests (Souza and Martins 2005). In addition to structural properties, the unique environmental pressures caused by seasonal flooding are known to have a pronounced influence on community composition, reflected in a high turnover between *terra firme* and *várzea* forests (Albernaz *et al.* 2012; Assis *et al.* 2014, 2015; Myster 2007, 2010) with levels of floristic similarity reaching just 10–30% (Wittmann *et al.* 2010).

Results from the Médio Juruá show significantly lower AGB in *várzea* forest than *terra firme*, demonstrating the overriding influence of the dramatic annual flood pulse. These differences in AGB were primarily driven by composition, rather than structural properties such as stem density or basal area (Hawes *et al.* 2012). Estimates based on allometric equations using dbh or basal area alone were actually similar or higher than in adjacent *terra firme* forest; lower AGB in *várzea* forest was only apparent when including stem-specific values for wood density and tree height, which depend upon composition and account for much of the variation in AGB estimates (Fearnside 1997; Nogueira *et al.* 2008). Despite the established hypothesis of lower wood density in dynamic environments such as Amazonian floodplains, based on the negative relationship with growth rates (Malhi *et al.* 2006), few *várzea* studies have successfully incorporated wood density which would ideally be measured *in situ* rather than relying on average values compiled across forest types.

Combining extensive and systematic sampling across the floodplain landscape in the Médio Juruá with detailed flood mapping allowed differences in AGB to be related to flood conditions including flood depth and duration (Hawes *et al.* 2012). Interestingly, the major finding here showed that, despite lower AGB in flooded forests than unflooded forests, AGB within *várzea* forest was lowest in plots flooded for the shortest annual period. This is possibly a result of more densely packed annual growth rings in areas with a shorter terrestrial phase and corresponding growth period (Wittmann *et al.* 2006; Worbes 1997; Worbes and Fichtler 2010), although if AGB actually peaks at intermediate flood duration then several other mechanisms may be important (Lucas *et al.* 2014). Models including landscape variables which are possible indicators of water stress, such as elevation, flood duration and distance to the nearest stream, consistently identified the binary distinction between forest types as the key factor (Hawes *et al.* 2012). This shows that, despite spatial and inter-annual variation in flood levels, there is a clear threshold between regularly flooded forests and the low-lying *terra firme* forests (including *paleo-várzeas*) that have remained unflooded for thousands of years, as supported by findings on community composition (Assis *et al.* 2014).

Temporal patterns of fruit availability, however, were generally similar in both *várzea* and *terra firme*, despite small differences in the phenology of leaves and flowers (Hawes and Peres 2016), consistent with typical fruiting in humid tropical forests during the early- to mid-rainy season (van Schaik *et al.* 1993; Zhang and Wang 1995). Fruiting peaked in *terra firme* forest during the mid-rainy season almost exclusively due to animal-dispersed plants, with wind-dispersed and ballistic genera bearing fruit at a more steady low level; in *várzea* forest, both wind- and water-dispersed genera bore fruits most frequently during the aquatic phase (Hawes and Peres 2016). There is evidence that *várzea* fruiting generally lags slightly behind that in *terra firme* (Haugaasen and Peres 2005), but also support for the possibility of a bimodal distribution in *várzea* fruit production, as first detected at Lago Teiú, Mamiraua (Ayres 1986). In the Médio Juruá, a secondary peak occurred during the terrestrial phase (onset of the rainy season), and was notably comprised of fleshy fruits such as *Byrsonima* spp. and *Manilkara* spp (Hawes and Peres 2016).

The proportion of plant genera exhibiting abiotic dispersal modes (wind, water and ballistic) was relatively higher in *várzea* forest (TF: 19 genera, 21.1% of genera; VZ: 17 genera, 26.2% of genera), whereas *terra firme* forest was more dominated by animal-dispersed plants (TF: 71 genera, 78.9% of genera; VZ: 48 genera, 73.8% of genera) (Hawes and Peres 2016). The relatively high prevalence of abiotic seed dispersal modes in *várzea* compared to *terra firme* was likely driven by differences in community composition, rather than differences within individual genera. The elevated proportion of wind-dispersed trees and lianas within *várzea* forest is also likely to be related to the lower stature and less continuous nature of the canopy, as well as the history of plant colonization of the floodplains from adjacent *terra firme* forest communities (Wittmann *et al.* 2010).

In light of other differences between the two forest types, the lack of a difference in total fruit biomass was surprising (Hawes and Peres 2016), since it implies that flooding and the subsequently higher nutrient availability in the floodplain forest (Furch 1997) does not affect fruit production. However, the simplified and fragmented *várzea* forest canopy suggests that relative fruit production (per stem or per unit of above-ground forest biomass) is actually higher in this forest type, where overall basal area is greater despite similar stem density (Hawes *et al.* 2012). That fruit biomass recorded in litterfall traps was similar, despite that these differences in forest structure is a testament to the high productivity of *várzea* forests.

The impact of the seasonal flood cycle is, however, clearly apparent within the animal community, including the resident frugivore assemblage (Ayres 1986; Haugaasen and Peres, 2005, 2008; Hawes and Peres 2014). For terrestrial vertebrates, such as caviomorph rodents, ungulates and ground-dwelling birds and reptiles, the barrier imposed by the floodwaters is absolute during the aquatic phase. These frugivores are understood to migrate to and from adjacent *terra firme* forests over the course of the year, returning as fruits and seeds fallen during the aquatic phase are exposed or deposited on the forest floor by the receding floodwaters (Haugaasen and Peres 2007), although these lateral migrations to track seasonal fruit resources are yet to be comprehensively explored. While terrestrial frugivores are effectively excluded from *várzea* forest during the flood pulse, the opposite scenario is the case for freshwater turtles (Balensiefer and Vogt 2006) and fish (Horn *et al.* 2011), including frugivorous characids and catfish, which abandon the river channel and oxbow lakes with the rising floodwaters to take advantage of canopy resources in *várzea* forests, including seeds, fruit pulp and arthropods (Goulding 1980). In contrast to terrestrial and aquatic species, most arboreal and scansorial vertebrates, including primates, squirrels and canopy birds, retain physical access to *várzea* forests all year round, although their relative abundances and diet may vary throughout the year between the two forest types (Hawes and Peres 2014).

The resultant binary networks produced contained low proportions of all potential interactions (TF: 25.7%, VZ: 19.4%) between 36 functional groups of frugivores and 152 plant genera (Hawes and Peres 2014). This results in a large number of diffuse interactions, which characterizes the typically low-specialization in tropical frugivore communities (Bascompte and Jordano 2007; Schleuning *et al.* 2012). Interestingly, our results suggested a greater degree of specialization in *várzea* forest, although further testing of weighted networks with greater coverage of rarely observed species is required to test this (Blüthgen *et al.* 2006, 2008). The large proportion of unobserved potential interactions in our networks also suggest a high probability of missing data due to sampling effects, in addition to truly "forbidden links", and highlight the need for further study to assess community-wide networks in such diverse fruit-frugivore communities. In particular, further work is still required to explore the potential overlap in the diet of frugivorous fish with other consumers (Correa *et al.* 2007; Goulding 1980; Horn *et al.* 2011; Kubitzki and Ziburski 1994). Such overlap could partly explain why recursive partitioning analysis failed to clearly match differences in fruit selection to fruit traits, despite finding significant heterogeneity in fruit resource use among broad frugivore guilds within each forest type (Hawes and Peres 2014).

4.4 Conclusion

Western Amazonian forests are of utmost importance in both global biodiversity and the global carbon balance, representing both a substantial source of emissions following deforestation and forest degradation, and a potential carbon sink if they can be adequately protected (Gibbs *et al.* 2007; Malhi *et al.* 2008). Floodplain forests such as *várzea* are particularly important in this regard, since they have been poorly mapped and assessed for carbon storage potential (Anderson *et al.* 2009; Hawes *et al.* 2012) and, as rivers represent the principal Amazonian transport routes, they face the highest pressure from rapidly expanding human settlements (Parolin *et al.* 2004a). Floodplain forests are also likely to bear the brunt of impacts from the development of new hydroelectric dams across the Amazon basin (Lees *et al.* 2016). As carbon stocks increasingly become the currency by which the success of conservation efforts are measured, it is necessary to assess the degree to which their preservation relates to the maintenance of both biodiversity and other important ecosystem services, including seed dispersal processes and animal protein provision to local residents.

Trait-based network approaches represent a useful tool to test the potential effects of overhunting on seed dispersal processes. Modeling the impact of a disproportionate loss of larger-bodied frugivores on the recruitment of large-seeded hardwood plant species allows the assessment of the more cryptic consequences of human disturbances on underlying ecosystem services, including carbon stocks (Bello *et al.* 2015; Peres *et al.* 2016). To date, this "empty forest" scenario has typically been examined in terrestrial habitats but, although much research is still needed to fill in the gaps in their extensive fruit-frugivore networks, *várzea* forests provide the potential setting to test the impacts of both overhunting and overfishing. The importance of imagining an empty flooded forest is further enhanced by the particular importance of fish for Amazonian livelihoods (Newton *et al.* 2011) and subsistence diets (Cerdeira *et al.* 1997). Demonstrating the role of vertebrate frugivores in the conservation of *várzea* forest, and *vice versa*, offers the possibility for the protection of both forest and its animal and human inhabitants. Increasing our understanding of this complex system of interactions and interdependence is therefore essential if:

1. the harvest of terrestrial or aquatic vertebrates is to be considered sustainable; and
2. the natural vegetation, and thus carbon stocks, of floodplain forests in Western Amazonia are to be preserved.

References

Albernaz, A.L., Pressey, R.L., Costa, L.R.F. *et al.* (2012) Tree species compositional change and conservation implications in the white-water flooded forests of the Brazilian Amazon. *Journal of Biogeography*, **39**, 869–883.

Anderson, L.O., Malhi, Y., Ladle, R.J. *et al.* (2009) Influence of landscape heterogeneity on spatial patterns of wood productivity, wood specific density and above ground biomass in Amazonia. *Biogeosciences*, **6**, 1883–1902.

Assis, R.L. and Wittmann, F. (2011) Forest structure and tree species composition of the understory of two central Amazonian *várzea* forests of contrasting flood heights. *Flora*, **206**, 251–260.

Assis, R.L., Wittmann, F., Piedade, M.T.F. and Haugaasen, T. (2014) Effects of hydroperiod and substrate properties on tree alpha diversity and composition in Amazonian floodplain forests. *Plant Ecology*, **216**, 41–54.

Assis, R.L., Haugaasen, T., Schöngart, J. *et al.* (2015) Patterns of tree diversity and composition in Amazonian floodplain paleo-*várzea* forest. *Journal of Vegetation Science*, **26**, 312–322.

Ayres, J.M. (1986) *Uakaris and Amazonian Flooded Forest*. PhD thesis. Cambridge University, Cambridge.

Baker, T.R., Phillips, O.L., Malhi, Y. *et al.* (2004) Variation in wood density determines spatial patterns in Amazonian forest biomass. *Global Change Biology*, **10**, 545–562.

Balensiefer, D.C. and Vogt, R.C. (2006) Diet of *Podocnemis unifilis* (Testudines, Podocnemididae) during the dry season in the Mamirauá Sustainable Development Reserve, Amazonas, Brazil. *Chelonian Conservation and Biology*, **5**, 312–317.

Balslev, H., Luteyn, J., Øllgaard, B. and Holm-Nielsen, L. (1987) Composition and structure of adjacent unflooded and floodplain forest in Amazonian Ecuador. *Opera Botamoca*, **92**, 35–57.

Bascompte, J. and Jordano, P. (2007) Plant-animal mutualistic networks: the architecture of biodiversity. *Annual Review of Ecology, Evolution, and Systematics*, **38**, 567–593.

Bello, C., Galetti, M., Pizo, M.A. *et al.* (2015) Defaunation affects carbon storage in tropical forests. *Science Advances*, **1**, e1501105–e1501105.

Blüthgen, N., Menzel, F. and Blüthgen, N. (2006) Measuring specialization in species interaction networks. *BMC Ecology*, **6**, 9.

Blüthgen, N., Fründ, J., Vázquez, D.P. and Menzel, F. (2008) What do interaction network metrics tell us about specialization and biological traits? *Ecology*, **89**, 3387–3399.

Bollen, A., Elsacker, L.V. and Ganzhorn, J.U. (2004) Relations between fruits and disperser assemblages in a Malagasy littoral forest: a community-level approach. *Journal of Tropical Ecology*, **20**, 599–612.

Campbell, D.G., Stone, J.L. and Rosas, J.R.A. (1992) A comparison of the phytosociology and dynamics of three floodplain (*várzea*) forests of known ages, Rio Jurua, western Brazilian Amazon. *Botanical Journal of the Linnean Society*, **108**, 213–237.

Cerdeira, R., Ruffino, M. and Isaac, V. (1997) Consumo de pescado e outros alimentos pela população ribeirinha do Lago Grande de Monte Alegre, PA. Brasil. *Acta Amazonica*, **27**, 213–227.

Comiskey, J., Dallmeier, F. and Aymard, G. (2000) Floristic composition and diversity of forested habitats in the Estación Biológica del Beni, Amazonian Bolivia, in *Biodiversity, Conservation and Management in the Region of the Beni*

Biological Station Biosphere Reserve, Bolivia (eds O. Herrera-MacBryde, F. Dallmeier, B. MacBryde *et al.*), Smithsonian Institution/Monitoring and Assessment of Biodiversity Program (SI/MAB), Rockville, MD, pp. 89–112.

Correa, S.B., Winemiller, K.O., López-Fernandez, H. and Galetti, M. (2007) Evolutionary perspectives on seed consumption and dispersal by fishes. *Bioscience*, **57**, 748–756.

Donatti, C.I., Guimarães, P.R., Galetti, M. *et al.* (2011) Analysis of a hyper-diverse seed dispersal network: modularity and underlying mechanisms. *Ecology Letters*, **14**, 773–81.

Fearnside, P.M. (1997) Wood density for estimating forest biomass in Brazilian Amazonia. *Forest Ecology and Management*, **90**, 59–87.

Fleming, T.H. and Kress, W.J. (2011) A brief history of fruits and frugivores. *Acta Oecologica*, **37**, 521–530.

Fleming, T.H., Breitwisch, R. and Whitesides, G.H. (1987) Patterns of tropical vertebrate frugivore diversity. *Annual Review of Ecology and Systematics*, **18**, 91–109.

Flörchinger, M., Braun, J., Böhning-Gaese, K. and Schaefer, H.M. (2010) Fruit size, crop mass, and plant height explain differential fruit choice of primates and birds. *Oecologia*, **164**, 151–61.

Freitas Alvarado, L. (1996) *Caracterizacion floristica y estructural de cuatro comunidades boscosas de la llanura aluvial inundable en la zona Jenaro Herrera, Amazonia Peruana. Documento Técnico No. 26*, Instituto de Investigaciones de la Amazonia Peruana (IIAP), Iquitos, Peru.

Furch, K. (1997) Chemistry of *varzea* and igapo soils and nutrient inventory of their floodplain forests, in *The Central Amazon Floodplain: Ecology of a pulsing system. Ecological Studies 126* (ed. W.J. Junk), Springer-Verlag, Berlin/Heidelberg, pp. 47–67.

Gautier-Hion, A., Duplantier, J.M., Quris, R. and Feer, F. (1985) Fruit characters as a basis of fruit choice and seed dispersal in a tropical forest vertebrate. *Oecologia*, **65**, 324–337.

Gentry, A.H. (1988) Tree species richness of upper Amazonian forests. *Proceedings of the National Academy of Sciences of the United States of America*, **85**, 156–159.

Gentry, A.H. (1996) *A Field Guide to the Families and Genera of Woody Plants of Northwest South America (Columbia, Ecuador, Peru)*, University of Chicago Press, Chicago, IL.

Gibbs, H.K., Brown, S., Niles, J.O. and Foley, J.A. (2007) Monitoring and estimating tropical forest carbon stocks: making REDD a reality. *Environ. Res. Lett.*, **2**, 45023.

Goulding, M. (1980) *The Fishes and the Forest: Explorations in Amazonian Natural History*, Univeristy of California Press, Berkeley, CL.

Griz, L.M.S. and Machado, I.C.S. (2001) Fruiting phenology and seed dispersal syndromes in caatinga, a tropical dry forest in the northeast of Brazil. *Journal of Tropical Ecology*, **17**, 303–321.

Haugaasen, T. and Peres, C.A. (2005) Tree phenology in adjacent Amazonian flooded and unflooded forests. *Biotropica*, **37**, 620–630.

Haugaasen, T. and Peres, C.A. (2007) Vertebrate responses to fruit production in Amazonian flooded and unflooded forests. *Biodiversity and Conservation*, **16**, 4165–4190.

Haugaasen, T. and Peres, C.A. (2008) Population abundance and biomass of large-bodied birds in Amazonian flooded and unflooded forests. *Bird Conserv. Int.*, **18**, 87–101.

Hawes, J.E. and Peres, C.A. (2014) Fruit–frugivore interactions in Amazonian seasonally flooded and unflooded forests. *Journal of Tropical Ecology*, **30**, 381–399.

Hawes, J.E. and Peres, C.A. (2016) Patterns of plant phenology in Amazonian seasonally flooded and unflooded forests. *Biotropica*, **48**, 465–475.

Hawes, J.E., Peres, C.A., Riley, L.B. and Hess, L.L. (2012) Landscape-scale variation in structure and biomass of Amazonian seasonally flooded and unflooded forests. *Forest Ecology and Management*, **281**, 163–176.

Hopkins, M.J.G. (2007) Modelling the known and unknown plant biodiversity of the Amazon Basin. *Journal of Biogeography*, **34**, 1400–1411.

Horn, M.H., Correa, S.B., Parolin, P. *et al.* (2011) Seed dispersal by fishes in tropical and temperate fresh waters: the growing evidence. *Acta Oecologica*, **37**, 561–577.

Irion, G., Junk, W.J. and Mello, J.A.S.N. (1997) The large central Amazonian river floodplains near Manaus: geological, climatological, hydrological and geomorphological aspects, in *The Central Amazon Floodplain Ecological Studies 126* (ed. W.F. Junk), Springer, Berlin Heidelberg, pp. 23–46.

Junk, W.J. (1989) Flood tolerance and tree distribution in Central Amazonian floodplains, in *Tropical Forests: Botanical dynamics, speciation, and diversity* (eds L.B. Holm-Nielsen, I.C. Nielsen and H. Balslev), Academic Press Limited, London, pp. 47–64.

Junk, W. J. (1997) General aspects of floodplain ecology with special reference to Amazonian floodplains. In: Junk, W. J. (ed.), *The Central Amazon Floodplain: Ecology of a pulsing system. Ecological Studies 126*, Springer-Verlag: Berlin/Heidelberg, pp. 3–20.

Junk, W.J., Piedade, M.T.F., Schöngart, J. *et al.* (2011) A classification of major naturally-occurring Amazonian lowland wetlands. *Wetlands*, **31**, 623–640.

Kitamura, S., Yumoto, T., Poonswad, P. *et al.* (2002) Interactions between fleshy fruits and frugivores in a tropical seasonal forest in Thailand. *Oecologia*, **133**, 559–572.

Kubitzki, K. and Ziburski, A. (1994) Seed dispersal in flood plain forests of Amazonia. *Biotropica*, **26**, 30–43.

Lees, A.C., Peres, C.A., Fearnside, P.M. *et al.* (2016) Hydropower and the future of Amazonian biodiversity. *Biodiversity and Conservation*, **25**, 451–466.

Levin, S.A., Muller-Landau, H.C., Nathan, R. and Chave, J. (2003) The ecology and evolution of seed dispersal: a theoretical perspective. *Annual Review of Ecology, Evolution, and Systematics*, **34**, 575–604.

Link, A. and Stevenson, P.R. (2004) Fruit dispersal syndromes in animal disseminated plants at Tinigua National Park, Colombia. *Revista Chilena de Historia Natural*, **77**, 319–334.

Lucas, C.M., Schöngart, J., Sheikh, P. *et al.* (2014) Effects of land-use and hydroperiod on above-ground biomass and productivity of secondary Amazonian floodplain forests. *Forest Ecology and Management*, **319**, 116–127.

Malhi, Y., Wood, D., Baker, T.R. *et al.* (2006) The regional variation of aboveground live biomass in old-growth Amazonian forests. *Global Change Biology*, **12**, 1107–1138.

Malhi, Y., Roberts, J.T., Betts, R.A. *et al.* (2008) Climate change, deforestation, and the fate of the Amazon. *Science*, **319**, 169–72.

Melack, J.M. and Hess, L.L. (2010) Remote sensing of the distribution and extent of wetlands in the Amazon basin, in *Amazonian Floodplain Forests: Ecophysiology, biodiversity and sustainable management* (eds W. Junk, M. Piedade, F. Wittmann *et al.*), Springer, Dordrecht/Berlin/Heidelberg/New York, pp. 43–60.

Myster, R.W. (2007) Interactive effects of flooding and forest gap formation on tree composition and abundance in the Peruvian Amazon. *Folia Geobotanica*, **42**, 1–9.

Myster, R.W. (2010) Flooding gradient and treefall gap interactive effects on plant community structure, richness, and alpha diversity in the Peruvian Amazon. *Ecotropica*, **16**, 43–49.

Myster, R.W. (2014) Interactive effects of flooding and treefall gap formation on *terra firme* forest and *várzea* forest seed and seedling mechanisms and tolerances in the Ecuadorean Amazon. *Community Ecology*, **15**, 212–221.

Nathan, R. and Muller-Landau, H.C. (2000) Spatial patterns of seed dispersal, their determinants and consequences for recruitment. *Trends in Ecology & Evolution*, **15**, 278–285.

Nebel, G., Kvist, L.P., Vanclay, J.K. *et al.* (2001) Structure and floristic composition of flood plain forests in the Peruvian Amazon: I: Overstorey. *Forest Ecology and Management*, **150**, 27–57.

Newton, P., Endo, W. and Peres, C.A. (2011) Determinants of livelihood strategy variation in two extractive reserves in Amazonian flooded and unflooded forests. *Environmental Conservation*, **39**, 97–110.

Nogueira, E., Nelson, B., Fearnside, P. *et al.* (2008) Tree height in Brazil's "arc of deforestation": shorter trees in south and southwest Amazonia imply lower biomass. *Forest Ecology and Management*, **255**, 2963–2972.

de Oliveira, A.C. and Piedade, M.T.F. (2002) Implicações ecológicas da fenologia reprodutica de *Salix martiana* Leyb. (Salicaceae) em áreas de várzea da Amazônia central. *Acta Amazonica*, **32**, 377–385.

Parolin, P. (2009) Submerged in darkness: adaptations to prolonged submergence by woody species of the Amazonian floodplains. *Annals of Botany*, **103**, 359–376.

Parolin, P., Ferreira, L.V., Albernaz, A.L.K.M. and Almeida, S.S. (2004a) Tree species distribution in *várzea* forests of Brazilian Amazonia. *Folia Geobotanica*, **39**, 371–383.

Parolin, P., de Simone, O., Haase, K. *et al.* (2004b) Central Amazonian floodplain forests: tree adaptations in a pulsing system. *Botanical Review*, **70**, 357–380.

Parolin, P., Lucas, C., Piedade, M.T.F. and Wittmann, F. (2010a) Drought responses of flood-tolerant trees in Amazonian floodplains. *Annals of Botany*, **105**, 129–139.

Parolin, P., Waldhoff, D. and Piedade, M.T.F. (2010b) Fruit and seed chemistry, biomass and dispersal, in *Amazonian Floodplain Forests: Ecophysiology, biodiversity and sustainable management* (eds W. Junk, M. Piedade, F. Wittmann *et al.*), Springer, Berlin/Heidelberg/New York, pp. 243–258.

Parolin, P., Wittmann, F. and Schöngart, J. (2010c) Tree phenology in Amazonian floodplain forests, in *Amazonian Floodplain Forests: Ecophysiology, biodiversity and sustainable management* (eds W. Junk, M. Piedade, F. Wittmann *et al.*), Springer, Berlin/Heidelberg/New York, pp. 105–126.

Parolin, P., Wittmann, F. and Ferreira, L.V. (2013) Fruit and seed dispersal in Amazonian floodplain trees – a review. *Ecotropica*, **19**, 19–36.

Peres, C.A. (2000) Effects of subsistence hunting on vertebrate community structure in Amazonian forests. *Conservation Biology*, **14**, 240–253.

Peres, C. and van Roosmalen, M.G.M. (2002) Primate frugivory in two species-rich Neotropical forests: implications for the demography of large-seeded plants in overhunted areas, in *Seed Dispersal and Frugivory: Ecology, evolution and conservation* (eds D.J. Levey, W.R. Silva and M. Galetti), CAB International, Wallingford, UK, pp. 407–422.

Peres, C.A., Emilio, T., Schietti, J. *et al.* (2016) Dispersal limitation induces long-term biomass collapse in overhunted Amazonian forests. *Proceedings of the National Academy of Sciences of the United States of America*, **113**, 892–897.

Pires, J.M. and Prance, G.T. (1985) The vegetation types of the Brazilian Amazon, in *Key Environments: Amazonia* (eds G.T. Prance and T.E. Lovejoy), Pergamon Press, Oxford, UK, pp. 109–145.

Prance, G.T. (1979) Notes on the vegetation of Amazonia III: the terminology of Amazonian forest types subject to inundation. *Brittonia*, **31**, 26–38.

Redford, K.H. (1992) The empty forest. *Bioscience*, **42**, 412–422.

van Roosmalen, M.G.M. (1985) *Fruits of the Guianan Flora*, Silvicultural Department of Wageningen Agricultural University.

van Schaik, C.P., Terborgh, J.W. and Wright, S.J. (1993) The phenology of tropical forests: adaptive significance and consequences for primary consumers. *Annual Review of Ecology and Systematics*, **24**, 353–377.

Schleuning, M., Fründ, J., Klein, A.-M. *et al.* (2012) Specialization of mutualistic interaction networks decreases toward tropical latitudes. *Curr. Biol.*, **22**, 1925–31.

Schöngart, J., Wittmann, F. and Worbes, M. (2010) Biomass and net primary production of central Amazonian floodplain forests, in *Amazonian Floodplain Forests: Ecophysiology, biodiversity and sustainable management* (eds W. Junk, M. Piedade, F. Wittmann *et al.*), Springer, Dordrecht/Berlin/Heidelberg/New York, pp. 347–388.

Sioli, H. (1984) The Amazon and its main affluents: hydrology, morphology of the river courses and river types, in *The Amazon: Limnology and the landscape ecology of a mighty tropical river and its basin* (ed. H. Sioli), Dr W. Junk Publishers, Dordrecht, pp. 127–166.

Smythe, N. (1986) Competition and resource partitioning in the guild of Neotropical terrestrial frugivorous mammals. *Annual Review of Ecology and Systematics*, **17**, 169–188.

Souza, A.F. and Martins, F.R. (2005) Spatial variation and dynamics of flooding, canopy openness, and structure in a Neotropical swamp forest. *Plant Ecol.*, **180**, 161–173.

Terborgh, J., Nunez-Iturri, G., Pitman, N.C.A. *et al.* (2008) Tree recruitment in an empty forest. *Ecology*, **89**, 1757–1768.

Voigt, F.A., Bleher, B., Fietz, J. *et al.* (2004) A comparison of morphological and chemical fruit traits between two sites with different frugivore assemblages. *Oecologia*, **141**, 94–104.

Wheelwright, N.T. (1985) Fruit-size, gape width, and the diets of fruit-eating birds. *Ecology*, **66**, 808–818.

Wilkie, D.S., Bennett, E.L., Peres, C.A. and Cunningham, A.A. (2011) The empty forest revisited. *Annals of the New York Academy of Sciences*, **1223**, 120–128.

Wittmann, F., Schöngart, J., Montero, J.C. *et al.* (2006) Tree species composition and diversity gradients in white-water forests across the Amazon Basin. *Journal of Biogeography*, **33**, 1334–1347.

Wittmann, F., Schöngart, J. and Junk, W.J. (2010) Phytogeography, species diversity, community structure and dynamics of central Amazonian floodplain forests, in *Amazonian Floodplain Forests: Ecophysiology, biodiversity and sustainable management* (eds W. Junk, M. Piedade, F. Wittmann *et al.*), Springer, Berlin/Heidelberg/New York, pp. 61–102.

Worbes, M. (1997) The forest ecosystem of the floodplains, in *The Central Amazon Floodplain: Ecology of a pulsing system. Ecological Studies 126* (ed. W.J. Junk), Springer-Verlag, Berlin/Heidelberg, pp. 223–266.

Worbes, M. and Fichtler, E. (2010) Wood anatomy and tree-ring structure and their importance for tropical dendrochronology, in *Amazonian Floodplain Forests: Ecophysiology, biodiversity and sustainable management* (eds W. Junk, M. Piedade, F. Wittmann *et al.*), Springer, Berlin/Heidelberg/New York, pp. 329–346.

Wright, S.J. (2003) The myriad consequences of hunting for vertebrates and plants in tropical forests. *Perspectives in Plant Ecology, Evolution and Systematics*, **6**, 73–86.

Wright, S.J., Andrés, H., Richard, C. *et al.* (2007) The bushmeat harvest alters seedling banks by favoring lianas, large seeds, and seeds dispersed by bats, birds, and wind. *Biotropica*, **39**, 363–371.

Zhang, S.-Y. and Wang, L.-X. (1995) Comparison of three fruit census methods in French Guiana. *Journal of Tropical Ecology*, **11**, 281–294.

5

Palm Diversity and Abundance in the Colombian Amazon

Henrik Balslev, Juan-Carlos Copete, Dennis Pedersen, Rodrigo Bernal, Gloria Galeano, Álvaro Duque, Juan Carlos Berrio and Mauricio Sanchéz

Abstract

Palms are majestic and emblematic in tropical forests where they sometimes form dense monodominat palm communities and at other times form communities with high richness of a variety of species and growth forms. Colombia's high palm species richness is associated with its geographical expansion and position in north-western South America at the cross-roads between Central and South America. This chapter gives detailed information concerning diversity and abundance of palms of the eastern Colombian Amazon near the frontiers of Brazil and Peru. The forest of the Colombian Amazon, known for its many palms, is a humid tropical forest. *Terra firme* forests in the Amazon harbor high diversities of palms. Floodplain and terrace forests had fewer species of palms compared to the *Terra firme* forest, but the palms were more abundant with 3,737 vs. 2,900 individuals per hectare on average.

Keywords *Floodplain forests; north-western South America; palm diversity; Terra firme forests; terrace forests; tropical forests*

5.1 Introduction

Palms are majestic and emblematic in tropical forests where they sometimes form dense monodominat palm communities and at other times form communities with high richness of a variety of species and growth forms. In Colombia, vegetation dominated by *Ceroxylon* palms are found in the Andes and in the Amazon, immense swamps dominated by *Mauritia flexuosa* are a common sight. On the coastal plain, *Euterpe oleracea* form large dense stands along the rivers. Palms are generally recognized as a group in contrast to many other plant families, probably most commonly so in the case of the coconut, *Cocos nucifera* or the introduced African oil palm, *Elaeis guineensis*. Palms may be the economically most important plant family for the inhabitants of the Amazon, who extract large amounts of a variety of palm products and use them for their daily activities and subsistence. Palms are used for constructing houses, for food, medicines and artisan activities and some palms are commercialized

Forest structure, function and dynamics in Western Amazonia, First Edition.
Edited by Randall W. Myster.

(Araújo and Lopes 2012; Balslev 2011; Balslev *et al.* 2008; Brokamp *et al.* 2011; Macía *et al.* 2011; Paniagua Zambrana *et al.* 2007).

Palms can be classified in eight growth forms, based on their morphology and size (Balslev *et al.* 2010). In the Colombian Amazon, most palms are cespitose with the stems branching near or just below the ground to form clumps and a smaller proportion of the palms are solitary. As far as the leaf shape goes, most palms in the Colombian Amazon have pinnate leaves and a smaller proportion have palmate or costapalmate leaves.

The Colombian palm flora is well described in the book *Palmas de Colombia* (Galeano and Bernal 2010). Colombia has 231 species of palms and it is the second richest American country in terms of palm species, only surpassed by Brazil with 273 species (Lorenzi *et al.* 2010). Colombia's high palm species richness is associated with its geographical expansion and position in north-western South America at the cross-roads between Central and South America. As far as palm biogeography is concerned, Colombia can be divided into four main regions:

1. The Colombian Amazon has the largest extension and reaches from the rocky outcrops of the Guianan Shield to the foothills of the Andes in the west and houses 92 palm species;
2. The Pacific lowlands include the Chocó region, which reaches from Panama to north-western Ecuador and is delimited by the Andes to the east; this region houses 86 species of palms;
3. The Andes region includes all areas of the cordillera above 500 m elevation and houses 82 species of palms; and
4. The Caribbean region, which includes the archipelagos of San Andrés and Providencia and is characterized by a very dry climate, includes only 14 species of palm (Galeano and Bernal 2010).

When it comes to the palm communities of Colombia they are much less studied than the taxonomic inventory, although there are a few studies in this respect (Ramírez-Moreno and Galeano 2011). This chapter gives detailed information concerning diversity and abundance of palms of the eastern Colombian Amazon near the frontiers of Brazil and Peru.

5.2 Study area

The Amazon covers 7,989,004 km^2 and is a mosaic of different forest types. It is estimated that the area has a flora of about 60,000 species and that it is one of the most diverse areas on the planet (Hoorn *et al.* 2010; ter Steege *et al.* 2003). The Colombian part of the Amazon covers 483,119 km^2, which corresponds to 42% of the area of the country. The forest of the Colombian Amazon, known for its many palms, is a humid tropical forest (Holdridge 1982; Holdridge *et al.* 1971). The climate is warm with an average annual temperature of about 25°C and average monthly precipitation of more than 100 mm and annual precipitation of 3,100–3,300 mm (Tobón 1999). There are two very different landscape types in the Colombian Amazon. One type is the *terra firme* landscapes which are

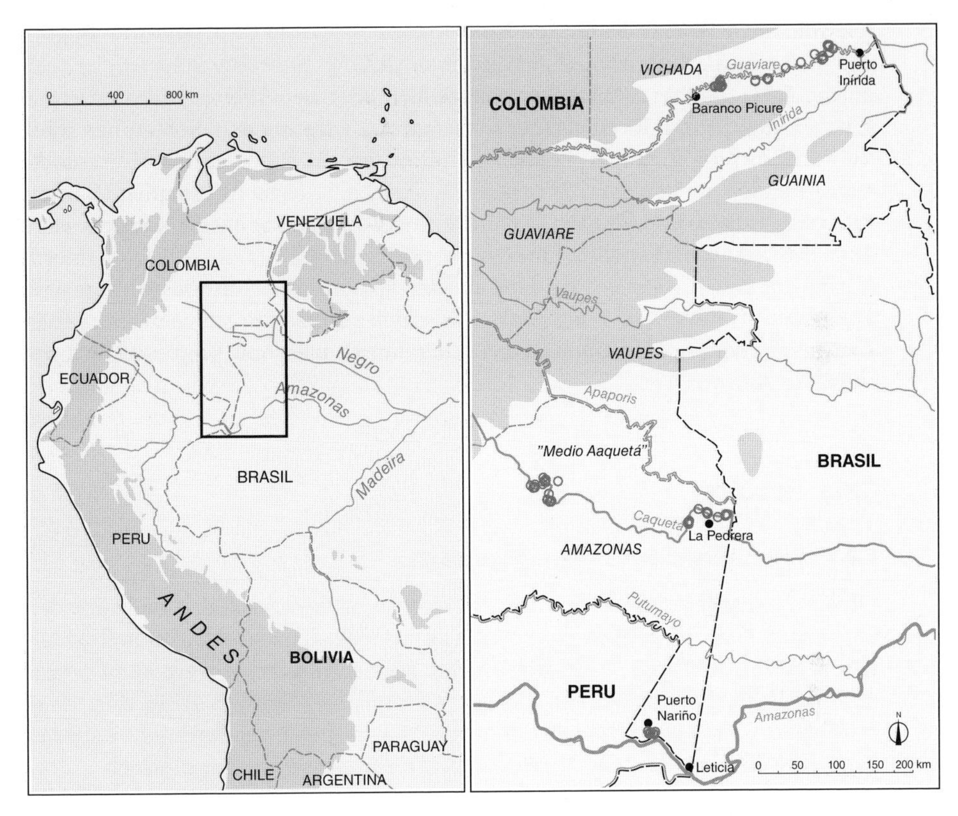

Figure 5.1 Study area in the eastern Colombian Amazon where 71 transects were placed along Río Guaviare (between the town of Inírida and the upstream settlement of Baranco Picure, 200 km to the west), Río Caquetá (from 42 km west to 11 km east of the village/military camp of La Pedrera), Medio Caqutá (225 km west of La Pedrera) and Río Amazonas (at Puerto Nariño, 68 km northwest of Leticia).

never flooded and generally have nutrient-poor and well-drained soils, with high local diversity. The other type is the floodplains and associated terraces which are flooded annually or even more rarely and also includes some ancient never-flooded terraces derived from rivers where they used to flow. Our transects were placed along the Guaviareas, Caquetá and Amazonas rivers (Figure 5.1).

5.3 Methods

We sampled palm data in 61 transects measuring 5 m × 500 m, following the procedure described by Balslev *et al.* (2010). The transects were subdivided into 100 subunits of 5 m × 5 m. In each subunit, all palm individuals were identified, counted and classified in one of four growth stage categories (seedling, juvenile, sub-adults, adult). Vouchers were taken whenever there was any doubt about the identity of the species and were deposited in the Herbarium (COL) of the

Instituto de Ciencias Naturales, Universidad Nacional de Colombia, Bogota and in the herbarium of Aarhus University (AAU). The collections can be seen on the AAU database (enter www.aubot.dk, collector = Bernal; number range 4,442–4,704). Along Río Guaviare we sampled 28 transects in September 2009, the team including Bernal, Pedersen and Balslev. At Río Caqueta 16 transects were sampled in January 2011, near La Pedrera, the team including Balslev, Pedersen, Copete and Nuñez. The 17 transects sampled at Medio Caqueta were done by Duque and Sanchez. Our dataset also includes 10 transects from Puerto Nariño on Río Amazonas sampled by Berrio with a slightly different design. These were transects of 4 m × 500 m, in which palms with stems of more than 2 cm were identified and counted. Details for all transects, locations, etc. are shown in Annex 5.1.

5.4 Results

5.4.1 Palms in *terra firme* forests (Figure 5.2)

Of the 71 transects, 38 (9 ha) were placed in *terra firme* forest and included 68 palm species in 20 genera and had an average density of 2,900 individuals per hectare (Table 5.1). The three most abundant species were *Oenocarpus bataua, Iriartella setigera* and *Oenocarpus bacaba* (Figure 5.3). In this forest type, small palms were the most common growth form, represented by species such as *Geonoma deversa, I. setigera* and *Lepidocaryum tenue,* which are all part of the understory. The palms that reach the canopy of the *terra firme* forest were *O. bataua, O. bacaba* and *Attalea maripa.* All six growth forms that we encountered were represented in the *terra firme* forests, where both the most common growth forms were small palms represented by 41 species and tall stemmed palms represented by 13 species, whereas other growth forms were represented by 1–5 species each (Figure 5.4). Of the 68 species in *terra firme* forests, 37 were cespitose and 26 were solitary, whereas the remaining 5 species were colonial (Figure 5.5). Leaves were mostly pinnate (65 species) and the exceptions were *Mauritiella armata* and *M. flexuosa* with costapalmate leaves and *L. tenue* with palmate leaves (Table 5.2).

5.4.2 Palms in floodplain and terrace forests (Figure 5.6)

We made 33 transects (8.25 ha) in floodplain forests and low-lying adjacent terraces. There we encountered 60 species of palms in 19 genera. The average density of palms in floodplains and terraces was 3,737 ind./ha (Table 5.1). The understory was dominated by small palms such as *Bactris fissifrons, Bactris macroacantha* and *Geonoma laxiflora.* The mid-story was dominated by medium-sized palms such as *Bactris riparia* and *Manicaria martiana* and the canopy was dominated by *Euterpe precatoria, Attalea butyracea* and *Astrocaryum aculeatum,* The most species-rich palm growth form in the floodplains and terraces were the small palms of which there were 35 species and the second-most common growth form was the large tall-stemmed palms of which there were 11 species, whereas other

Annex 5.1 Localities for the 71 transects in which palms were inventoried. Transect number were given in the field and used in the transect database kept at Aarhus University. Habitat is given as FP = floodplain or TF = *terra firme*. Alt is elevation above sea level. X and Y are geographic coordinates recorded with GPS in the field at the beginning of each transect.

Transect HB#	Dept.	Province	Locality	Habitat	Alt (m)	X	Y
1000	Vichada	Río Guaviare	46 km W of Inírida	FP	113	3° 59’43.38"	68°19’16.82"
1001	Vichada	Río Guaviare	46 km W of Inírida	FP	98	3°59’15.03"	68°19’17.62"
1002	Guainía	Río Guaviare	40 km W of Inírida	FP	100	3°57’57.85"	68°16’25.47"
1003	Guainía	Río Guaviare	43 km W of Inírida	FP	113	3°55’18.25"	68°18’57.44"
1004	Vichada	Río Guaviare	46 km W of Inírida	FP	110	3°59’51.24"	68°19’44.12"
1005	Vichada	Río Guaviare	49 km W of Inírida	FP	109	4°00’26.36"	68°20’21.62"
1006	Vichada	Río Guaviare	46 km W of Inírida	FP	113	4°00’19.62"	68°19’13.17"
1007	Guainía	Río Guaviare	52 km W of Inírida	FP	113	3°52’57.52"	68°23’47.34"
1008	Guainía	Río Guaviare	51 km W of Inírida	FP	116	3°51’41.64"	68°23’07.17"
1009	Guainía	Río Guaviare	52 km W of Inírida	FP	116	3°50’36.97"	68°24’02.56"
1010	Vichada	Río Guaviare	62 km W of Inírida	FP	116	3°54’11.16"	68°29’22.17"
1011	Vichada	Río Guaviare	82 km W of Inírida	FP	116	3°48’44.42"	68°39’03.53"
1012	Vichada	Río Guaviare	100 km W of Inírida	FP	111	3°44’52.13"	68°49’44.53"
1013	Guainía	Río Guaviare	126 km W de Inírida	TF	152	3°38’08.53"	69°02’48.08"
1014	Guainía	Río Guaviare	126 km W of Inírida	TF	150	3°37’32.82"	69°02’26.01"
1015	Guainía	Río Guaviare	125 km W of Inírida	TF	156	3°37’12.09"	69°02’02.59"
1016	Guainía	Río Guaviare	125 km W of Inírida	FP	121	3°37’28.32"	69°01’50.55"
1017	Guainía	Río Guaviare	141 km W of Inírida	FP	125	3°36’06.99"	69°10’50.03"

(*continued*)

Annex 5.1 (Continued)

Transect HB#	Dept.	Province	Locality	Habitat	Alt (m)	X	Y
1018	Guainía	Río Guaviare	142 km W of Inírida	FP	118	3°36’26.98"	69°10’51.15"
1019	Guainía	Río Guaviare	185 km W of Inírida	TF	163	3°33’25.82"	69°34’10.64"
1020	Guainía	Río Guaviare	184 km W of Inírida	TF	155	3°33’11.03"	69°33’42.19"
1021	Guainía	Río Guaviare	186 km W of Inírida	TF	179	3°32’54.11"	69°34’38.17"
1022	Guainía	Río Guaviare	186 km W of Inírida	TF	176	3°32’10.76"	69°34’39.11"
1023	Guainía	Río Guaviare	188 km W of Inírida	TF	171	3°33’29.31"	69°36’01.21"
1024	Guainía	Río Guaviare	189 km W of Inírida	TF	172	3°32’47.85"	69°36’27.03"
1025	Guainía	Río Guaviare	192 km W of Inírida	TF	176	3°32’55.74"	69°38’03.61"
1026	Guainía	Río Guaviare	193 km W of Inírida	TF	145	3°32’23.75"	69°38’39.55"
1027	Guainía	Río Guaviare	185 km W of Inírida	FP	128	3°35’06.39"	69°34’32.26"
1028	Vichada	Río Guaviare	186 km W of Inírida	FP	132	3°36’11.26"	69°34’58.20"
1029	Caquetá	La Pedrera	8 km NE of La Pedrera	FP	80	1°16’50.35"	69°31’05.97"
1030	Caquetá	La Pedrera	8 km NE of La Pedrera	FP	85	1°17’00.85"	69°30’38.17"
1031	Caquetá	La Pedrera	8 km NE of La Pedrera	FP	81	1°16’58.98"	69°30’18.92"
1032	Caquetá	La Pedrera	6 km E of La Pedrera	FP	83	1°17’54.14"	69°31’18.90"
1033	Caquetá	La Pedrera	2 km W of La Pedrera	TF	120	1°18’49.38"	69° 36’31.02"
1034	Caquetá	Loma Linda	17 km W of La Pedreda	TF	120	1° 15’58.12"	69° 43’21.90"
1035	Caquetá	Loma Linda	18 km W of La Pedrera	TF	105	1° 15’23.31"	69° 44’15.83"
1036	Caquetá	Loma Linda	18 km W of La Pedrera	FP	80	1° 15’0369"	69° 45’03.49"
1037	Caquetá	Loma Linda	13 km W of La Pedrera	TF	88	1° 13’19.81"	69° 49’53.51"

1038	Caquetá	Loma Linda	17 km N of La Pedrera	TF	109	1° 15’48.08"	69° 43’43.50"
1039	Caquetá	Loma Linda	19 km W of La Pedrera	TF	85	1° 16’34.03"	69° 43’42.08"
1040	Caquetá	Los Ingleses	41 km W of La Pedrera	TF	105	1° 22’39.95"	69° 57’26.47"
1041	Caquetá	Los Ingleses	40 km NE of La Pedrera	TF	103	1° 21’19.35"	69° 56’24.81"
1042	Caquetá	Los Ingleses	40 km NE of La Pedrera	TF	104	1° 21’53.68"	69° 56’54.80"
1043	Caquetá	Los Ingleses	40 km NE of La Pedrera	TF	119	1° 23’26.22"	69° 56’43.04"
1044	Caquetá	Los Ingleses	40 km NE of La Pedrera	TF	119	1° 23’11.15"	69° 57’06.48"
1045	Caquetá	Rìo Caquetá	229 km SE of La Pedrera	TF	133	0°54’26.28"	71°27’33.31"
1046	Caquetá	Río Caquetá	231 km SE of La Pedrera	TF	139	0°54’3.24"	71°36’55.80"
1047	Caquetá	Río Caquetá	230 km SE of La Pedrera	TF	108	0°53’16.80"	71°36’11.52"
1048	Caquetá	Río Caquetá	233 km SE of La Pedrera	TF	149	0°52’1.20"	71°37’51.24"
1049	Caquetá	Río Caquetá	223 km SE of La Pedrera	FP	116	1° 6’59.04"	71°34’55.20
1050	Caquetá	Río Caquetá	224 km SE of La Pedrera	FP	141	1° 6’54.36	71°35’12.12"
1051	Caquetá	Río Caquetá	244 km SE of La Pedrera	FP	149	0°58’59.88	71°44’50.64
1052	Caquetá	Río Caquetá	242 km SE of La Pedrera	FP	131	0°58’30.36	71°43’41.16"
1053	Caquetá	Río Caquetá	242 km SE of La Pedrera	TF	116	0°58’19.56"	71°43’41.16"
1054	Caquetá	Río Caquetá	248 km SE of La Pedrera	TF	118	0°56’39.48"	71°43’41.16"
1055	Caquetá	Río Caquetá	248 km SE of La Pedrera	TF	113	0°57’16.92"	71°46’46.20"
1056	Caquetá	Río Caquetá	229 km SE of La Pedrera	FP	111	0°55’6.96	71°36’16.92"
1057	Caquetá	Río Caquetá	232 km SE of La Pedrera	FP	114	0°56’23.28"	71°38’10.32"

(*continued*)

Annex 5.1 (Continued)

Transect HB#	Dept.	Province	Locality	Habitat	Alt (m)	X	Y
1058	Caquetá	Río Caquetá	223 km SE of La Pedrera	FP	100	1° 2'50.64"	71°33'58.68"
1059	Caquetá	Río Caquetá	218 km SE of La Pedrera	FP	105	1° 7'40.08"	71°32'9.96"
1060	Caquetá	Río Caquetá	219 km SE of La Pedrera	FP	108	1° 8'11.04"	71°32'25.44"
1061	Caquetá	Río Caquetá	221 km SE of La Pedrera	FP	100	1° 8'11.04"	71°33'31.68"
1062	Amazonas	Puerto Nariño	3 km NW of Puerto Nariño	TF	94	3°45'16"	70° 24'07.38"
1063	Amazonas	Puerto Nariño	2 km NW of Puerto Nariño	TF	107	3°45'25.91"	70° 23'15.12"
1064	Amazonas	Puerto Nariño	4 km NW of Puerto Nariño	TF	104	3°45'07.80"	70° 23'09.04"
1065	Amazonas	Puerto Nariño	2.5 km NW of Puerto Nariño	TF	112	3°44'58.32"	70° 23'02.50"
1066	Amazonas	Puerto Nariño	3.5 km NW of Puerto Nariño	TF	113	3°44'24.02"	70° 22'56.93"
1067	Amazonas	Puerto Nariño	4 km NE of Puerto Nariño	TF	104	3°45'25.76"	70° 21'016.31"
1068	Amazonas	Puerto Nariño	3 km NE of Puerto Nariño	TF	131	3°43'41.11"	70° 24'48.95"
1069	Amazonas	Puerto Nariño	2.8 km E of Puerto Nariño	TF	94	3°44'01.67"	70° 19'48.31"
1070	Amazonas	Puerto Nariño	5 km E of Puerto Nariño	TF	104	3°45'45.07"	70° 20'28.83"
1071	Amazonas	Puerto Nariño	4.5 km NE of Puerto Nariño	TF	103	3°45'11.46"	70° 21'27.45"

Figure 5.2 Some common canopy (a, b, c) and understory (d, e, f) palm species in the *terra firme* forest of eastern Colombian Amazon. (a) *Astrocaryum chambira*. (b) *O. bataua*. (c) *I. deltoidea*. (d) *I. setigera*. (e) *Geonoma maxima* var maxima. (f) *Bactris simplicifrons*.

Table 5.1 Diversity and abundance of palms in eastern Colombian Amazon, studied in 71 transects along the Guaviare, Caquetá and Amazonas rivers.

	Transects	Individuals/ ha	Species	Species/ Transect (Median)	Fisher's alpha	Growth Forms	Architecture
Río Caquetá	33	3075	61		8,62	6	3
Río Guaviare	28	4347	27		4,25	6	3
Río Amazonas	10	565	23	7–11(9)	3,62	5	2
Terra firme	38	2900	67	6–26(16)	9,6	7	3
Floodplain	33	3737	60	3–28(15)	8,31	7	3
Total	71	3301	73	3–28(15)		7	3

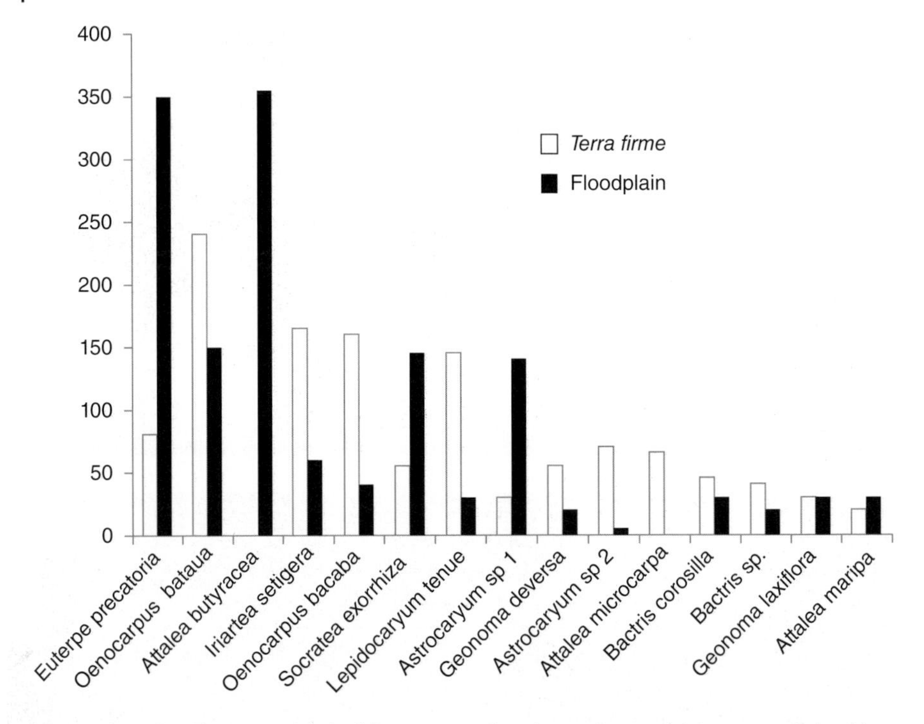

Figure 5.3 Abundances (ind./ha) of the 15 most abundant palm species in eastern Colombian Amazon, recorded in 71 transects (17.25 ha) in *terra firme* and floodplain forest.

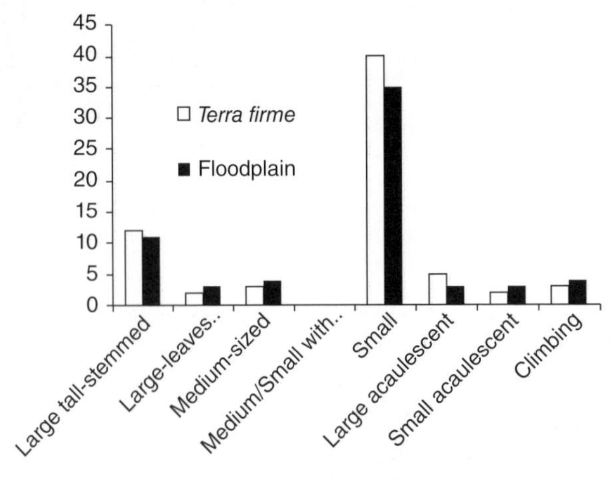

Figure 5.4 Numbers of species of palms in eastern Colombian Amazon in each growth form, as defined in Balslev *et al.* (2011) and in each habitat (*terra firme*/floodplain and terraces).

growth forms all had less than 5 species (Figure 5.4). Of the 60 species, 32 were cespitose, 22 were solitary and 6 were colonial (Figure 5.5). The dominant leaf form of the palms in the floodplain and terrace forest was the pinnate, which was found in 55 species. Three species had costaplamate leaves (*M. flexuosa,*

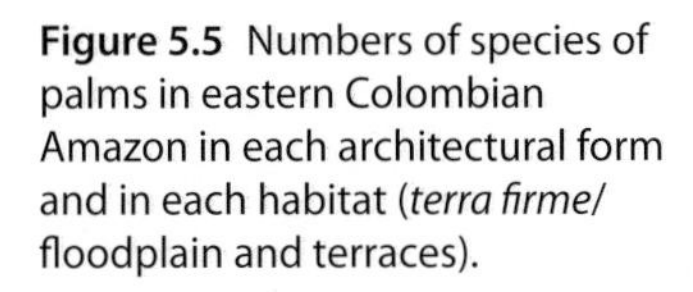

Figure 5.5 Numbers of species of palms in eastern Colombian Amazon in each architectural form and in each habitat (*terra firme*/ floodplain and terraces).

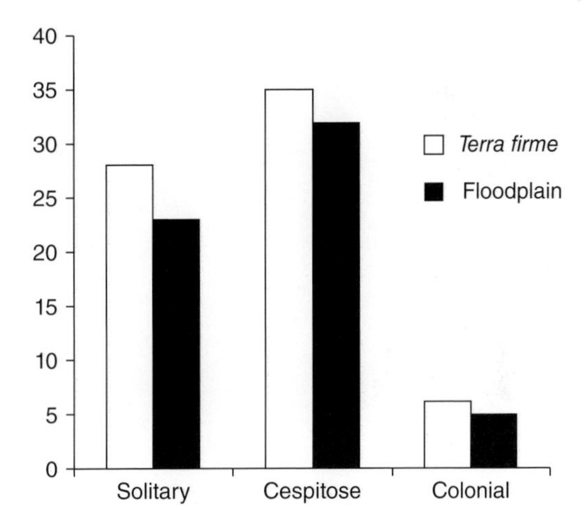

Table 5.2 Number of palm species in 71 transects in the eastern Colombian Amazon with pinnate, palmate, and costapalmate leaves, respectively.

Leaf shape	*Terra firme*	Floodplain and terraces	Total
Pinnate	65	55	69
Palmate	1	2	2
Costapalmate	2	3	3
Total	68	60	74

Mauritiella aculeate and *M. armata*) and one species had palmate leaves (*Chelyocarpus ulei*) (Table 5.2).

5.4.2.1 Growth forms

Of the eight growth forms defined for tropical American palms (Table 5.3; Balslev *et al.* 2011), we found seven in our sample, the only missing growth form being Medium/Small Palms with Stout Stems. The most common growth form was that of Small Palms, which was represented by 43 (58%) of the species, whereas large Tall-stemmed Palms were represented by 13 (18%) of the species, and the remaining growth forms with less than 5 species each (Annex 5.2). The two habitat types (*terra firme*; floodplains and terraces) were almost identical in their composition in terms of number of palm species in each growth form (Figure 5.4). The number of growth forms encountered in the three parts of our study area were 6 in Río Guaviare and Río Caquetá and 4 in Río Amazonas (Table 5.1).

5.4.2.2 Palm architecture

We encountered three architectural types among the 74 species of palms in our sample. The most common form was that of cespitose palms, which was found in 39 (53%) of the species, whereas solitary had 29 (39%) and colonial had only 6 (8%) of the species (Annex 5.2). Again the different architectural types were

Figure 5.6 Some common canopy (a, b, c, d) and understory (e, f, g) palm species from the floodplain and terrace forests of eastern Colombian Amazon. (a) *Mauriatia flexuosa*. (b) *Astrocaryum jauri*. (c) *A. butyracea*. (d) *E. precatoria*. (e) *Bactris major var major*. (f) *Desmoncus polyacanthos*. (g) *Manicaria saccifera*.

Table 5.3 Growth forms of American palms defined by overall size of stems and leaves, whether caulescent or acaulescent, and whether climbers or not (table based on Balslev *et al.* 2011).

Palm Growth form	Stem height (m)	Stem diam. (cm)	Leaf size (m)	Stem development	Self-supporting/ Climbing
1. Large Tall-stemmed	20–35	20–100	2.5–10 (+)	Caulescent	Self-supporting
2. Large-leaved Medium-Short-stemmed	1–20	15–25	4–10	Caulescent/ Acaulescent	Self-supporting
3. Medium-sized	8–15	12–15	2–4	Caulescent	Self-supporting
4. Medium/smal with Stout Stems	1–20	30–60	2–4	Caulescent	Self-supporting
5. Small	0.8–8	0.4–12	0.2–2.5	Caulescent	Self-supporting
6. Large Acaulescent	n.a.	n.a.	4–8	Acaulescent	Self-supporting
7. Small Acaulescent	n.a.	n.a.	1–2	Acaulescent	Self-supporting
8. Climbing	4–30	0.5–2	1–2	Caulescent	Climbing

equally distributed across the two main habitats (*terra firme*/floodplains and terraces) in our study area (Figure 5.5). The three different palm architectures were all represented in our study sites at Río Guaviare and Río Caquetá, but only two were represented in our sample at Río Amazonas.

5.4.2.3 Palm species richness (Table 5.1)

Overall we encountered 73 palm species (Annex 5.2) in 20 genera in our 71 transects that covered 17.25 hectares of *terra firme* and floodplain/terrace forest. The *terra firme* forests were the richest with 67 species, whereas the floodplain/terrace forests had 60 species of palms. Comparing the three sites included in our study, the Río Caqueta is by far the richest, having representation of 61 of the 73 species encountered overall. Río Guaviare has less than half the number of species (27) and finally our site at Río Amazonas is the least rich with 23 species.

5.4.2.4 Palm diversity

As for species richness, Fischer's alpha was higher in *terra firme* compared floodplain/terrace forests. Río Cauetá had a very high score on the Fischer's alpha (8.62) and Río Guaviare with less than half the number of species (27) also had a much lower Fischer's alpha (4.25). Finally, our site at Río Amazonas is the least rich with 23 species and Fischer's alpha of 3.62.

5.4.2.5 Palm abundance

We encountered a total of 56,937 individual palms in the 71 transects covering 17.25 hectares, giving an average density of 3,301 individuals per hectare. The

Annex 5.2 Palm species found in 61 transects of 5 × 500 m (15.25 ha) and 10 transects of 4 × 500 m (2 ha) in eastern Colombian Amazon arranged according to their total abundance in this study (column 3; Total ind. all transects). Vouchers were collected in the number series of Rodrigo Bernal (RB) and are deposited in the herbarium (COL) of Instituto de Ciencias, Universidad Nacional de Colombia. Columns 4 and 5 give the total number of individuals collected in *terra firme* and floodplain forest transects, respectively. Palm architecture refers to whether the palms were solitary (sol), cespitose (ces) or colonial (col). Growth form follows the classification proposed by Balslev *et al.* 2011 (1 = Large Tall-Stemmed Palms, 2 = Large-Leaved Medium-short Stemmed Palms, 3 = Medium-sized Palms, 4 = Medium/Small Palms with Stout Stems, 5 = Small Palms, 6 = Large Acaullescent Palms, 7 = Small Acaulescent Palms, and 8 = Climbing Palms). Leaf shape is given as pinnate (pin), palmate (pal) or costa-palmate cos (cop).

Species Name	Voucher collection RB #	Total ind. all transects	Terra firme forest	Floodplain forest	Palm architecture	Growth form	Leaf shape
1. *Euterpe precatoria* Mart.	4,475	7,457	1,393	6,064	sol	1	pin
2. *Oenocarpus bataua* Mart.	4,452	6,759	4,055	2,704	sol	1	pin
3. *Attalea butyracea* (Mutis ex L.f.) Wess. Boer	4,451	6,144		6,144	sol	1	pin
4. *Iriartella setigera* (Mart.) H. Wendl.	4,462	4,073	2,927	1,146	col	5	pin
5. *Oenocarpus bacaba* Mart.	4,449	3,538	2,833	705	sol	1	pin
6. *Socratea exorrhiza* (Mart.) H. Wendl.	4,442	3,447	961	2,486	sol	1	pin
7. *Lepidocaryum tenue* Mart.		2,902	2,481	421	col	5	pal
8. *Astrocaryum ciliatum* F. Kahn and B. Millán		2,343	399	1,944	sol	6	pin
9. *Geonoma deversa* (Poit.) Kunth.	4,466	1,662	1,160	502	ces	5	pin
10. *Astrocaryum gynacanthum* Mart.	4,521	1,603	1,493	110	ces	5	pin
11. *Attalea microcarpa* Mart.		1,424	1,424		sol	6	pin
12. *Bactris corossilla* H. Karst.	4,524	1,368	864	504	ces	5	pin
13. *Bactris acanthocarpa* Mart.	4,447	1,152	819	333	ces	5	pin
14. *Geonoma laxiflora* Mart.	4,687	1,082	514	568	ces	5	pin
15. *Attalea maripa* (Aubl.) Mart.	4,474	943	369	574	sol	1	pin

Annex 5.2 (Continued)

Species Name	Voucher collection RB #	Total ind. all transects	Terra firme forest	Floodplain forest	Palm architecture	Growth form	Leaf shape
16. *Bactris macroacantha* Mart.	4,694	914	144	770	ces	5	pin
17. *Geonoma stricta* (Poit.) Kunth var. *stricta*	4,672	878	393	485	ces	5	pin
18. *Bactris hirta* Mart. var. *hirta*	4,468	770	368	402	ces	5	pin
19. *Bactris simplicifrons* Mart.	4,443	755	398	357	ces	5	pin
20. *Bactris major* Jacq. var. *major*	4,453	566		566	ces	5	pin
21. *Bactris fissifrons* Mart.		682	258	424	ces	5	pin
22. *Bactris maraja* Mart. var. *maraja*	4,490	464	239	225	ces	5	pin
23. *Geonoma maxima* (Poit.) Kunth var. *maxima*	4,446	427	174	253	ces	5	pin
24 . *Astrocaryum* sp		424	424		sol	1	pin
25. *Bactris martiana* A.J. Hend.	4,695	412	2	410	ces	5	pin
26. *Attalea racemosa* Spruce		344	37	307	sol	6	pin
27. *Phytelephas macrocarpa.* Ruiz and Pav.		308	308		sol	2	pin
28. *Manicaria saccifera* Gaertn.	4,519	283		283	ces	2	pin
29. *Geonoma oligoclona* Trail	4,678	270	140	130	ces	5	pin
30. *Geonoma leptospadix* Trail		225	55	170	ces	5	pin
31. *Geonoma camana* Trail	4,701	253	9	244	sol	5	pin
32. *Bactris brongniartii* Mart.	4,445	235	1	234	col	5	pin
33. *Bactris killipii* Burret	4,681	233	121	112	ces	5	pin

(*continued*)

Annex 5.2 (Continued)

Species Name	Voucher collection RB #	Total ind. all transects	Terra firme forest	Floodplain forest	Palm architecture	Growth form	Leaf shape
34. *Desmoncus polyacanthos* Mart.	4,495	214	92	122	ces	8	pin
35. *Mauritia flexuosa* L. f.		210	1	209	sol	1	cop
36. *Astrocaryum jauari* Mart.	4,450	190	187	3	ces	1	pin
37. *Geonoma maxima* var. *spixiana* (Mart.) A.J. Hend.		187	116	71	ces	5	pin
38. *Hyospathe elegans* Mart.	4,515	191	150	41	ces	5	pin
39. *Astrocaryum aculeatum* G. Mey.		159	23	136	sol	1	pin
40. *Attalea insignis* (Mart.) Drude		156	46	110	sol	6	pin
41. *Oenocarpus balickii* F. Kahn		151	121	30	sol	3	pin
42. *Geonoma macrostachys* var. *acaulis* (Mart.) A.J. Hend.		123	65	58	sol	7	pin
43. *Bactris hirta* var. *lakoi* (Burret) A.J. Hend.		118	57	61	ces	5	pin
44. *Oenocarpus minor* Mart.		114	14	100	ces	3	pin
45. *Geonoma macrostachys* Mart. var. *macrostachys*	4,444	106	47	59	sol	7	pin
46. *Geonoma stricta* var. *trailii* (Burret) A.J. Hend.	4,670	77	47	30	ces	5	pin
47. *Geonoma brongniartii* Mart.		63	12	51	sol	5	pin
48. *Iriartea deltoidea* Ruiz and Pav.		58	26	32	sol	1	pin
49. *Desmoncus mitis* Mart. var. *mitis*	4,461	55	23	32	ces	8	pin
50. *Astrocaryum chambira* Burret.		162	124	38	sol	1	pin

Annex 5.2 (Continued)

Species Name	Voucher collection RB #	Total ind. all transects	Terra firme forest	Floodplain forest	Palm architecture	Growth form	Leaf shape
51. *Bactris balanophora* Spruce		53	53		ces	5	pin
52. *Bactris riparia* Mart.		34	1	33	col	5	pin
53. *Chamaedorea pauciflora* Mart.		33	30	3	sol	5	pin
54. *Bactris halmoorei* A.J. Hend.	4,671	25	15	10	ces	5	pin
55. *Attalea plowmanii* (Glassman) Zona		23	23		sol	6	pin
56. *Chelyocarpus ulei* Dammer		14		14	sol	5	pin
57. *Desmoncus mitis* var. *leptospadix* (Mart.) A.J. Hend.	4,517	13	11	2	ces	8	pin
58. *Bactris gasipaes* Kunth		11	11		ces	1	pin
59. *Bactris acanthocarpa* var. *exscapa* Barb. Rodr.		12	12		ces	5	pin
60. *Bactris hirta* Mart. var. *pectinata*	4,473	6	5	1	ces	5	pin
61. *Bactris* sp.		6	6		ces	5	pin
62. *Manicaria martiana* Burret	4,685	5	5		ces	2	pin
63. *Desmoncus giganteus* A.J. Hend.	4,529	7	6	1	ces	8	pin
64. *Mauritiella armata* (Mart.) Burret	4,469	4	1	3	ces	3	cop
65. *Aiphanes deltoidea* Burret	4,677	3	3		sol	5	pin
66. *Geonoma maxima* var. *chelidonura* (Spruce) A.J. Hend.		3	2	1	ces	5	pin
67. *Oenocarpus circumtextus* Mart.	4,683	2	2		sol	5	pin
68. *Syagrus smithii* (H.E. Moore) Glassman		2	1	1	sol	5	pin

(*continued*)

Annex 5.2 (Continued)

Species Name	Voucher collection RB #	Total ind. all transects	Terra firme forest	Floodplain forest	Palm architecture	Growth form	Leaf shape
69. *Wettinia drudei* (O.F. Cook and Doyle) A.J. Hend.		2	1	1	col	5	pin
70. *Astrocaryum macrocalyx* Burret		1		1	sol	2	pin
71. *Bactris sphaerocarpa* Trail		1	1		ces	5	pin
72. *Chamaedorea pinnatifrons* (Jacq.) Oerst.		1	1		sol	5	pin
73. *Mauritiella aculeata* (Kunth) Burret		1		1	ces	5	cop
74. *Wettinia augusta* Poepp. ex Endl.		1	1		col	5	pin

most abundant species was *E. precatoria* followed by a group of 7 species that all had an average of over 100 individuals per hectare (*O. bataua, Attalea butyracea, I. setigera, O. bacaba, Socratea exorrhiza, Astrocaryum ciliatum* and *L. tenue*). The remaining species represent a total of increasingly less abundant species, including 19 species with less than one individual per hectare. The rarest species in our sample were only encountered once and accordingly are very rare in our study sites (*Astrocaryum macrocalyx, Bactris sphaerocarpa, Chamaedorea pinnatifrons, M. aculeata* and *Wettinia augusta*) (Annex 5.2), although they all have medium to large overall ranges and sometimes are common elsewhere.

5.4.2.6 Palm leaf shape

Of the 74 species encountered in our study, 69 had pinnate leaves, 3 had costapalmate leaves (*M. flexuosa, M. aculeata* and *M. armata*) and 2 had palmate leaves (*C. ulei* and *L. tenue*) (Table 5.3; Annex 5.2). The three leaf shapes were present in both habitats (*terra firme*/floodplain and terrace) in similar proportions. The three leaf shapes were all present in our study areas along Río Guaviare and Río Caqueta, but only two were found near Río Amazonas at Puerto Nariño (Table 5.3; Annex 5.2).

5.5 Discussion

Terra firme forests in the Amazon harbor high diversities of palms. In an area of the western Amazon that included parts of Bolivia, Peru, Ecuador and Colombia, there were 77 species of palms (Kristiansen *et al.* 2012). In the present study, we found a high richness of palms for this kind of forest and it also included seven out of the eight growth forms that have been proposed for tropical American palms, and three architectural types. These results agree with the findings of Balslev *et al.* (2012) when they studied the sub-Andean forests in Bolivia. In that study, the *terra firme* forests were the most species rich. This type of forest is in many respects similar to the floodplain forests, and we recorded similar numbers of species, the same growth forms and architectural types, even if there was a difference in abundance of palms, the floodplain forest having more abundant palm communities. Our results are similar to those of Montufar and Pintaud (2006), who studied the composition, abundance and microhabitat preferences of 52 species of palms near Jenaro Herrera and 41 species near Intuto in the Amazon of Peru.

Floodplain and terrace forests had fewer species of palms compared to the *terra firme* forest, but the palms were more abundant with 3,737 *vs.* 2,900 individuals per hectare on average (Table 5.1). Our results for species richness in floodplain forest are much higher than those reported by Kahn and Mejia (1990), who stated that the inundated forest had low palm diversity but with high palm abundance. Our species richness is also higher than what was reported for the floodplain forest along Río Ucayali in Peru (Balslev *et al.* 2010), who found only 18 species and abundances of 1,460 palms per hectare. In Bolivia, the floodplain forests had only 17 species and 3,400 individuals per hectare in this kind of forest.

The palms have different growth forms, which makes it possible for them to occupy different niches in the forest. In tropical America, Small Palms are the most common growth form and it is occupied by 423 species in 36 genera. Small Palms dominate the understory of the forests. Large Tall-stemmed Palms, the next most common growth form, which is occupied by 102 species in 19 genera, are common in the forest canopy, at least their adult individuals, whereas their seedlings and juveniles share the under-story with the smaller growth forms (Balslev *et al.* 2011). In this study the growth form Small Palms were also the most common and were represented by the highest number of species in both *terra firme* and in floodplain/terrace forests. This shows that the under-story is the most diverse part of the forests as far as palms are concerned. Our results are corroborated by the studies of tropical American forest in general (reviewed in Balslev *et al.* 2011) and also in a subsequent study in Bolivia (Balslev *et al.* 2012).

The cespitose architecture is in general more common in Amazonian palms and that was also found in this study, although solitary palms are represented by almost as many species. The colonial growth form, in contrast, is less well-represented. The cespitose architecture has the obvious advantage that it allows for both vegetative and sexual regeneration, whereas the solitary architecture depends entirely on sexual reproduction involving production of seeds and their subsequent germination and establishment. The dominance of cespitose species (53%) in our study is similar to what have been found elsewhere in the Amazon (Balslev *et al.* 2010).

The species richness of palms encountered in this study is high, and the *terra firme* forests were just a bit richer (67 species) than the forests on floodplains and low terraces (60 species). The rarefaction curve (Figure 5.7) shows significant differences between the three areas within our study site. The Río Caqueta

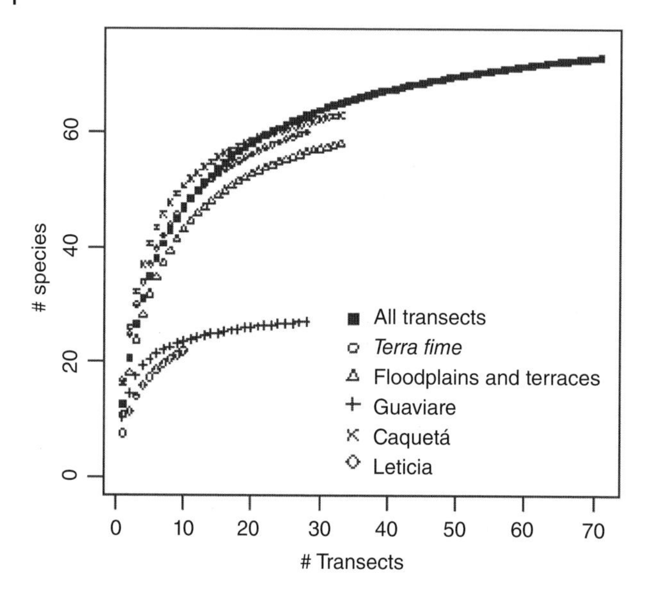

Figure 5.7 Rarefaction curve for palm species richness in 71 transects in eastern Colombian Amazon, total and divided by habitat (*terra firme*/floodplain) and for the three groups of transects made at the Guaviare, Caquetá and Amazonas rivers.

region is very rich in species, whereas Río Guaviare and Río Amazonas both have comparatively lower species richness. We found a total of 73 species in 21 genera. In comparison, the palm flora in the central Amazon included 26 species in 9 genera in one study, and another study in the western Amazon in the Ucayali basin reported 29 species in 19 genera (Kahn and de Granville 1992). At the Yasuni station, there were 34 species in 16 genera (Svenning 1999). In a study of the western Amazon, there were 48 species (Vormisto *et al.* 2004). In eastern Peru, two studies at Jenaro Herrera and Intuto reported 52 and 41 palm species (Montufar and Pintaud 2006). Along the upper Río Ucayali, there were 51 palm species (Balslev *et al.* 2010) and in the Bolivian lowlands of the southern Amazon, there were 38 species (Balslev *et al.* 2012), whereas in a broader study in the western Amazon, covering plots from Bolivia to Colombia, there were 95–98 species of palms (Kristiansen *et al.* 2011, 2012). This high palm species richness is caused by environmental conditions including topography, hydrology, soils and climate (Eiserhardt *et al.* 2011; Henderson 1995).

The Amazonian forests are also known for their abundance of palms and some of the palms are part of the group of species that makes the forests oligarchic, that is, dominated by a few very abundant and wide-ranging species (Macía and Svenning 2005). The oligarchic nature of the forest depends, apart from the palms, on very abundant species in the families Fabaceae, Violaceae, Moraceae, Meliaceae, Rubiaceae, Bombacaceae and Euphorbiaceae. This study confirms the palms as being part of the Amazonian forest oligarchies. As we registered all individuals of palms in our transects, we document here the super-abundance of *E. precatoria*, followed by *O. bataua*, *A. butyracea*, *I. setigera*, *O. bacaba*, *S. exorrhiza* and *A. ciliatum*. It is noteworthy that these are the same palms that are super abundant in southern Amazonia in Bolivia (Balslev *et al.* 2012). In Bolivia, however, there were some other species that also attain super-abundance and are wide-ranging, such as *G. deversa*, the most abundant, followed by *Iriartea deltoidea, Geonoma occidentalis, O. bataua, S. exorrhiza, Astrocaryum gratum, Hyospathe elegans, Dictyocaryum lamarckianum* and *E. precatoria*.

The most common leaf shape among palms in the eastern Colombian Amazon is the pinnate leaf, which was found in 68 of the 73 species. The exceptions were two species with palmate leaves (*L. tenue* and *C. ulei*), and costapalmate in three species (*M. flexuosa, M. armata* and *M. aculeata*). The dominance of pinnate leaves reflect the biogeographic fact that the South American continent is dominated by the palm subfamily Arecoideae and to some degree Ceroxyloideae, both with pinnate leaves, whereas the subfamily Coryphoideae, which has palmate leaves and the subfamily Calamoideae with costapalmate leaves, are almost absent (Balslev *et al.* 2010). This dominance of Arecoideae and Ceroxyloidea is different from the Caribbean and Central American forests, where the subfamily Coryphoideae is common and palmate leaves consequently dominate the palm communities. In Yucatan, for instance, in the forests of Quintana Ro, 5 of 11 species have palmate leaves (Alvarado-Segura *et al.* 2012). The dominance of pinnate leaves was also documented for the palm flora of upper Río Ucayali, where all palms except four species had pinnate leaves (Balslev *et al.* 2010), and the forests of the Bolivian Amazon, where all species had pinnate leaves except the widespread *M. flexuosa* with costapalmate leaves.

Acknowledgements

We are greatful to the European Commission's FP7 program (Contract no. 213126 to HB) for funding our research. We thank all who contributed with information and help in the field, particularly the communities along Río Guaviare, Río Caquetá and Río Amazonas.

References

Alvarado-Segura, A.A., Calvo-Irabién, L.M., Duno de Stefano, R. and Balslev, H. (2012) Palm species richness, abundance and diversity in the Yucatan Peninsula, in a Neotropical context. *Nordic Journal of Botany*, **30**, 613–622.

Araújo, F.R. and Lopes, M.A. (2012) Diversity of use and local knowledge of palms (Arecaceae) in eastern Amazonia. *Biodiversity and Conservation*, **21**, 487–501.

Balslev, H. (2011) Palm harvest impacts in north-western South America. *The Botanical Review*, **77**, 370–380.

Balslev, H., Grandez, C., Paniagua-Zambrana, N.Y. *et al.* (2008) Palmas (Arecaceae) útiles en los alrededores de Iquitos, Amazonía Peruana. *Revista Peruana de Biología*, **15** (suppl. 1), 121–132.

Balslev, H., Eiserhardt, W., Kristiansen, T. *et al.* (2010) Palms and palm communities in the upper Ucayali River valley – a little-known region in the Amazon basin. *Palms*, **54** (2), 57–72.

Balslev, H., Kahn, F., Millan, B. *et al.* (2011) Species diversity and growth forms in tropical American palm communities. *The Botanical Review*, **77**, 381–425.

Balslev, H., Pérez Durán, Z., Pedersen, D. *et al.* (2012) Subandean and adjacent lowland palm communities in Bolivia. *Ecología en Bolivia*, **47** (1), 7–36.

Brokamp, G., Valderrama, N., Mittelbach, M. *et al.* (2011) Trade in palm products in north-western South America. *The Botanical Review*, **77**, 571–606.

Eiserhardt, W.L., Svenning, J.-C., Kissling, W.D. and Balslev, H. (2011) Geographical ecology of the palms (Arecaceae): determinants of diversity and distributions across spatial scales. *Annals of Botany*, **108**, 1391–1416.

Galeano, G. and Bernal, R. (2010) *Palmas de Colombia* Guía de campo. Editorial Universidad Nacional de Colombia. Instituto de Ciencias Naturales. Facultad de Ciencias, Universidad Nacional de Colombia, Bogotá, p. 628.

Henderson, A. (1995) *The Palms of the Amazon*, Oxford University Press, New York.

Holdridge, L.R. (1982) *Ecología basada en zonas de vida*, Instituto Interamericano de Ciencias Agricoles, San José.

Holdridge, L.R., Grenke, W.C., Hatheway, W.H. *et al.* (1971) *Forest Environmenst in Tropical Life Zones: A pilot study*, Pergamon Press, Oxford. UK.

Hoorn, C., Wesselingh, F.P., ter Steege, H. *et al.* (2010) Amazonia through time: Andean uplift, climate change, landscape evolution, and biodiversity. *Science*, **330**, 927–931.

Kahn, F. and de Granville, J.-J. (1992) *Palms in Forest Ecosystems of Amazonia. (Ecological studies 95)*, Springer-Verlag, Berlin.

Kahn, F. and Mejia, K. (1990) Palm communities in wetland forest ecosystems of Peruvian Amazonia. *Forest Ecology and Management* **33**/34, 169–179.

Kristiansen, T., Svenning, J.-C., Pedersen, D. *et al.* (2011) Local and regional palm (Arecaceae) species richness patterns and their cross-scale determinants in the western Amazon. *Journal of Ecology*, **99**, 1001–1015.

Kristiansen, T., Svenning, J.-C., Eiserhardt, W.L. *et al.* (2012) Environment versus dispersal in the assembly of western Amazonian palm communities. *Journal of Biogeography*, **39**, 1318–1332.

Lorenzi, H., Noblik, L.R., Kahn, F. and Ferreira, E. (2010) *Flora Brasileira, Arecaceae (Palmeiras)*, Instituto Plantarum de Estudios da Flora, Ltda., Nova Odessa.

Macía, M.J. and Svenning, J.-C. (2005) Oligarchic dominance in western Amazonian plant communities. *Journal of Tropical Ecology*, **21**, 613–626.

Macía, M.J., Armesilla, P.J., Cámara-Leret, R. *et al.* (2011) Palm uses in northwestern South America: a quantitative review. *The Botanical Review*, **77**, 462–570.

Montufar, R. and Pintaud, J.-C. (2006) Variation in species composition, abundance and microhabitat preferences among western Amazonian *terra firme* palm communities. *Botanical Journal of the Linnean Society*, **151**, 127–140.

Svenning, J.-C. (1999) Microhabitat specialization in a species-rich palm community in Amazonian Ecuador. *Journal of Ecology*, **87**, 55–56.

Paniagua Zambrana, N.Y., Byg, A., Svenning, J.-C. *et al.* (2007) Diversity of palm uses in the western Amazon. *Biodiversity and Conservation*, **16**, 2771–2787.

Ramírez-Moreno, G. and Galeano, G. (2011) Comunidades de palmas en dos bosques de Chocó, Colombia. *Caldasia*, **32** (2), 315–329.

ter Steege, H., Pitman, N., Sabatier, D. *et al.* (2003) A spatial model of tree a – diversity and tree density for the Amazon. *Biodiversity and Conservation*, **12**, 2255–2277.

Tobón, M.C. (1999) *Monitoring and modeling hydrological fluxes in support of nutrient cycling studies in Amazonian rain forest ecosystems* Dissertation, University of Amsterdam, Amsterdam.

Vormisto, J., Svenning, J.-C., Hall, P. and Balslev, H. (2004) Diversity and dominance in palm (Arecaceae) communities in terra firme forests in the western Amazon basin. *Journal of Ecology*, **92**, 577–588.

6

Why Rivers Make the Difference: A Review on the Phytogeography of Forested Floodplains in the Amazon Basin

Florian Wittmann and Ethan Householder

Abstract

River wetlands have been an important landscape component of the Amazon Basin since the existence of neotropical rainforests. The modern Amazon landscape and its geomorphological history are intricately tied to its river wetlands and the ways in which they have reworked its surface. This chapter describes the principal ways in which river wetlands alter the Amazonian landscape. It outlines how plants respond to the altered conditions and discusses how these processes play out on the Amazonian landscape at ecological and evolutionary scales. Wetlands have been the major evolutionary force in vascular plant diversification, particularly in tropical regions. The Miocene marks an important period in the evolutionary history of the Amazonian landscape and its wetlands, largely due to Andean mountain building. Throughout the Quaternary, global sea-level fluctuations affected the Amazon river and its major tributaries as far inland as 2,500 km from the Atlantic coast.

Keywords *Amazon Basin; Amazon landscape; Andean mountain; Atlantic coast; global sea-level fluctuations; neotropical rainforests; plant diversification; Quaternary; river wetlands; tropical regions*

6.1 Introduction

While the relevance of river wetlands to biogeography is globally recognized, little attention has been given to their role in plant biogeography in the Amazon Basin. This is particularly surprising as the "Amazon wetscape" drains the largest volume of water, includes the largest freshwater wetlands, and houses the most diverse freshwater plant community on Earth. River wetlands have been an important landscape component of the Amazon Basin since the existence of neotropical rainforests. The modern Amazon Basin is covered by approximately 600,000 km^2 of river wetlands (7–8% of its total area), and a third of its area is covered by ancient alluvial substrates. Thus, the modern Amazon landscape and its geomorphological history are intricately tied to its river wetlands and the ways in which they have reworked its surface.

Forest structure, function and dynamics in Western Amazonia, First Edition.
Edited by Randall W. Myster.

In this chapter we describe the principal ways in which river wetlands alter the Amazonian landscape, we outline how plants respond to the altered conditions and we discuss how these processes play out on the Amazonian landscape at ecological and evolutionary scales. Because Amazonian river wetlands are mostly forested, we address these questions with emphasis on trees. In face of current threats to Amazonian river wetlands by humans and global climate change, we also call attention to the urgent need for conservation measures of these globally unique wetland ecosystems.

6.2 The geological history of flood-pulsing wetlands in the Amazon Basin

6.2.1 Through the Paleogene

Wetlands have been the major evolutionary force in vascular plant diversification, particularly in tropical regions (Greb et al. 2006). The northern part of South America has been characterized by a hot and humid climate since the Upper Jurassic, at least 152 million years BP. Although there is no evidence for moist tropical floras in northern South America for much of the Cretaceous, pollen of tropical mangroves in Colombia, Venezuela and the Guianas suggest the existence of warm, equable climates at low altitudes at approximately 59 Ma BP in the Upper Paleocene (resumed in Burnham and Johnson 2004). During the Paleocene–Eocene transition, pollen data suggest a transition from a flora dominated by taxa of pantropical elements to a flora dominated by taxa of neotropical elements after the boundary (Rull 1999). Both pollen data and macrofossils undeniably confirm the existence of modern neotropical rainforests in large parts of northern South America since the Upper Eocene, approximately 50 Ma BP (Burnham and Graham 1999; Burnham and Johnson 2004).

6.2.2 The Miocene

The Miocene marks an important period in the evolutionary history of the Amazonian landscape and its wetlands, largely due to Andean mountain building. Perhaps most importantly, Andean mountain building profoundly changed the amount and seasonality of rain falling on the Amazon Basin (Hoorn *et al.* 2010). By establishing a monsoon climate, the Andes produced a highly seasonal water influx into the lowlands (Irion and Kalliola 2010; Kaandorp *et al.* 2005). The strong annual flood pulse would increasingly regulate species interaction and ecological function of an evolving Amazonian wetland system.

Prior to Andean uplift, the "pan-Amazonian" region was characterized by a widespread river system, including the area of the present Amazon, Orinoco and Magdalena drainage systems, and extended southwards to the Paraná Basin (Hoorn *et al.* 2010). This ancient river system was fundamentally changed with Andean uplift, which began during the Paleogene, but was most intense during the Miocene from 23 to 10 Ma BP. Specifically, Andean uplift generated a system of depressions in the foreland basin of the western Amazon, giving rise to two vast freshwater wetland systems, the Pebas (23–10 Ma BP) and Acre (10 Ma BP). Although many details are still debated, a series of marine transgressions likely repeatedly invaded. More certain however, is that these wetlands dominated an area of several tens to hundreds of thousands kilometers squared and dominated much of Western Amazonia for upwards of 10 million years (Hoorn *et al.* 2010).

During this time, erosion of the newly emerging highlands continued to fundamentally change the sedimentary landscape of the Amazon. Andean-derived sediments were trapped in the low-lying Pebas and Acre system until approximately 10 Ma BP, after which they began to reach the Atlantic Ocean at about 7 Ma BP (Figueiredo *et al.* 2010). By this time, Western Amazonia had developed into a landscape characterized by entrenched rivers with high sediment load and widespread river terrace systems (Hoorn *et al.* 2010; Irion *et al.* 1997). Fluvial deposition of eroded Andean sediments blanketed an extensive area, extending from the Andean base as a triangle towards the central part of the Amazon Basin near the city of Manaus. The Caquetá-Japurá river to the north, and the Madeira river to the south form the approximate natural boundaries of the Andean sediment in the west, and older, cratonic geological formations of central and eastern Amazonia.

The extensive fluvial deposit laid down during the Miocene forms much of what we know today as the *terra firme* of Western Amazonia. This alluvial blanket is associated with highest tree species diversity relative to other parts of the basin. The younger and more nutrient-rich alluvial soils have been linked to demographic plant traits that speed growth, shorten lifespan, increase turnover rate and decrease generation time, thus providing an ecological basis for higher diversification and species richness in Western Amazonia (Baker *et al.* 2014). Other authors considering the high species diversity of Western Amazonian forests postulate that bouts of allopatric speciation were caused by the vast Pebas and Acre wetlands that fragmented upland forests (Antonelli and Sanmartín 2011). The draining of these wetlands during the late Miocene would have also opened huge areas of *terra firme* habitat for recolonization, presumably spurring diversification as well (Fine *et al.* 2014). In all of these potential scenarios, Miocene wetlands of Western Amazonia play the pivotal role in generating its exceptionally high species richness.

6.2.3 The Quaternary

Throughout the Quaternary, global sea-level fluctuations affected the Amazon river and its major tributaries as far inland as 2,500 km from the Atlantic coast (Irion *et al.* 2010). Floodplains increased in area when sea levels rose, such as during the late Pleistocene (~120,000 years BP). In contrast, when sea levels were low, rivers deeply incised previously deposited sediment (e.g. during the last glacial maximum, 21,000 years BP). Relicts of river incision are still found in the hundreds of finger-shaped, open water bodies, or ria lakes, that characterize the lower sections of some Amazonian rivers in the central part of the basin. Similar lakes in the western Amazon do not exist because topographical depressions have been rapidly filled in by rivers with large sediment loads (Irion *et al.* 2010). Today they form part of the *várzea*, a floodplain habitat type occurring along sediment-laden rivers. The current extent of modern *várzea* was formed by the backwater effect of higher sea levels during the Holocene, when large parts of the Amazon valley were submerged. Remnants of even higher sea levels exist as large river terraces, or paleo-*várzeas* that mostly formed during the

Sangamonian age, approximately 110,000 years BP (Irion 1984; Irion *et al.* 2010, "fossil floodplains" *sensu* Räsänen *et al.* 1987). Although these terraces are older and 15–20 m higher than modern *várzea*, paleo-*várzeas* have a distinctly *várzea*-like mineral composition (Irion 1984). These occur in most parts of the western basin and along the Amazon main stem and some tributaries in central Amazonia. There are also relicts of older Pleistocene *várzea* terraces, but these show a higher degree of mineral alteration compared to those of Sangamonian *várzea* (Irion 1984). Much of the paleo-*várzea* is no longer flooded, in which case it is regarded under an umbrella term as *terra firme*.

Alternating cycles of dryer/hotter and wetter/cooler climates are expected to have characterized the Quaternary. While Quaternary climate change is likely to have affected water discharge of Amazonian rivers and the extents of their respective floodplains, the volume of water flowing through the Amazon riverscape was too large for major rivers to dry out. Thus, climate change would not have interrupted interactions between wetlands and flora and fauna. Throughout the Quaternary, rivers are even thought to have imposed dispersal barriers to many upland vertebrate species, making a strong case for their large size and landscape persistence (Silva *et al.* 2005). For trees, Amazonian rivers may have served as a refuge from drought during postulated dry periods – wetlands effectively buffer against drought because of moist edaphic conditions (Wittmann *et al.* 2006; 2010; 2013).

This brief history of the Amazon Basin shows that freshwater rivers have been continuous components of the Amazonian landscape since at least the existence of the modern neotropical flora (Wittmann *et al.* 2013). With exception of the Orinoco floodplains, presumably no other freshwater floodplain system was uninterruptedly present in any other place in the world (however, little data are available for the African *Cuvette Centrale* of the Congo River basin, another equatorial, flood-pulsing and forested river system (Betbeder *et al.* 2014; Evrard 1968)). Stable conditions over evolutionary time are likely to have presented ample opportunity for adaptation and the development of a specialized floodplain flora. It is therefore not surprising that many Amazonian wetland trees can tolerate submergence of some or all plant organs for up to 300 days year^{-1}. Both the high diversity of Amazonian floodplain forests – the most diverse worldwide (Wittmann *et al.* 2006) – and their relatively high number of habitat specialists (Wittmann *et al.* 2013) are strong indications that Amazonian wetland trees have developed a diversity of strategies to cope with the specific conditions of their environment.

6.3 Floodplain environments: why rivers make the difference

6.3.1 Trees and flooding

Flooding represents a complex of stressors that impose several concurrent challenges to plant functioning (Jackson and Colmer 2005). Inundations cause hypoxic or anoxic soil conditions within a few hours. Respiring roots and microorganisms rapidly deplete oxygen supplies (Armstrong *et al.* 1994; Crawford 1989, 1992; Visser *et al.* 2003), which are not quickly restored because oxygen diffuses poorly through water and submerged plant tissue. Sediment deposits of flood waters may additionally deteriorate soil aeration, and thus oxygen supply. Oxygen depletion is accompanied by increased levels of CO_2, anaerobic decomposition of organic matter, increased solubility of mineral substances, and reduction of the soil redox potential (Joly and Crawford 1982; Kozlowski and Pallardy 1997). Many potentially toxic compounds can also accumulate due to altered composition of the soil micro-flora (Parolin 2009). Finally, when floodwaters are turbid, shoots of completely submerged trees cannot photosynthesize due to the lack of sunlight (Jackson and Colmer 2005; Parolin 2009).

Despite heavy constraints imposed by flooding, many Amazonian tree species are highly productive in flood-prone areas (Wittmann *et al.* 2010). Amazonian tree species subject to flooding compensate oxygen deficiency through biochemical, physiological and morpho-anatomical adaptations. Many trees that tolerate long-term flooding reduce their metabolism during waterlogging, resulting in decreased photosynthetic rates, and reduced wood and shoot growth (Fernandez *et al.* 1999; Schöngart *et al.* 2002; Worbes 1986). Further tree adaptations to flooding include partial or complete leaf shedding with the onset of flooding, adventitious roots and/or specialized roots such as pneumatophores, increased root biomass, hypertrophic lenticels on roots and stems, root aerenchyma, resprouting, transition to fermentative enzymes under anaerobic conditions, and elevated levels of anti-oxidants (e.g. De Simone *et al.* 2002; Ferreira *et al.* 2007; Oliveira Wittmann *et al.* 2007; Parolin 2009; Parolin *et al.* 2002; Schlüter *et al.* 1993; Wittmann and Parolin 2005; Worbes 1997).

Differences in the ecophysiological response of trees to waterlogging generally lead to a well-defined species zonation along flooding gradients (Ferreira 1997; Wittmann *et al.* 2002). Thus, the composition and richness of tree species is strongly influenced by differences in flood height and duration, which reach up to 10 m and 300 days $year^{-1}$. Under such extreme conditions, inundation represents the single-most limiting factor that controls tree species establishment, distribution and growth (Junk *et al.* 1989).

Most specialist tree species of the Amazonian floodplains are thought to be ecotypes originating from the surrounding *terra firme*, a notion mostly based on floristic resemblance between these two habitats (Kubitzki 1989; Wittmann *et al.* 2010). The number of generalist tree species that floodplains have in common with adjacent *terra firme* can be high, especially where floods are short and ephemeral (Terborgh and Andresen 1998; Wittmann and Junk 2003). However, it is this strong overlap that is likely to have provided ample opportunity for experimentation of trees with flooded habitat. Over time this may have provided traction for selective processes to operate along flood gradients, promoting niche differentiation of flood-tolerant ecotypes and floodplain specialists (Wittmann *et al.* 2013). The idea that floodplain specialists evolve from upland species thus rests on the assumption of strong tradeoffs between species' flood tolerance and their ability to compete in non-flooded *terra firme* (Wittmann *et al.* 2010).

The high number of endemic and specialist tree species in Amazonian floodplains has long been noted (Mori 2001; Prance 1979). Examining the distribution patterns of over 600 floodplain species, Wittmann *et al.* (2013) estimated that endemic habitat specialists (geographically and ecologically restricted) might account for between 10 and 33%, depending on the region and the type of floodplain. Molecular studies are also shedding new light, suggesting than many cryptic floodplain species or races may remain yet to be recognized. For example, in a study of *Himatanthus sucuuba* (Apocynaceae), populations from floodplains and *terra firme* habitats showed strong genetic differentiation; however, these molecular differences were not accompanied by any taxonomically relevant morphological variation (Ferreira *et al.* 2009; 2010).

6.3.2 Trees and dispersal in semi-aquatic habitats

In many Amazonian floodplain tree species, seed dispersal mechanisms, seed morphology and seedling emergence are strongly regulated by their interaction with the annual flood-pulse and aquatic organisms (Oliveira Wittmann *et al.* 2007; Parolin 2000). For example, many floodplain tree species synchronize fruiting with high water levels (Kubitzki and Ziburski 1994; Wittmann and Parolin 1999), increasing the changes that seeds will be dispersed by water. Water dispersal or hydrochory – in general an uncommon phenomenon in trees – can be observed in several floodplain species. Seed buoyancy is a key adaptation for water dispersal: seeds can be kept afloat by air-containing tissues in fruits and seeds for periods of up to two months (Kubitzki 1985; Kubitzki and Ziburski 1994; Waldhoff *et al.* 1996; Ziburski 1991). Contact with river water is thought to be the most important factor breaking seed dormancy in some Amazonian floodplain tree species (Scarano 1998; Ziburski 1991). However, for non-floating seeds, longer dormancy periods are needed to prevent seedlings from exposure to a submerged and hypoxic environment, unsuitable for germination (Frankland *et al.* 1987; Kozlowski and Pallardy 1997). In these species, germination is timed to the terrestrial phase of the flood pulse (Oliveira Wittmann *et al.* 2010).

Fish also may play an important role in seed dispersal and reproductive dynamics of many Amazonian floodplain specialist tree species (Correa *et al.* 2015; Gottsberger 1978). Many floodplain trees produce fleshy fruits that attract fruit-eating fish (Kubitzki 1985) and fruiting phenology is often synchronized with annual foraging patterns, thus increasing the chances of dispersal. Fish dispersal is potentially advantageous, because it can move seeds against the prevailing water current (Anderson *et al.* 2005) and increase dispersal distance, especially in non-buoyant seeds (Correa *et al.* 2007). Furthermore, seed passage through the gut plays a major role in breaking seed dormancy (Ziburski 1991) and can increase germination rates (Mannheimer *et al.* 2003).

6.3.3 Trees and alluvial soils

The single-most important factor determining alluvial substrate conditions is the underlying geology of floodplain catchment areas. Based on geological criteria, two predominant classes of freshwater floodplain types exist within the Amazon Basin: white-water *várzeas,* and black- or clear-water *igapós* (Junk *et al.* 2011;

Sioli 1954). The *várzeas* drain the Andes or the Andean foothills. They include the floodplains of the Solimões–Amazon main stem, the Juruá, Japurá, Purús and Madeira rivers, and collectively cover approximately 400,000 km^2 (Melack and Hess 2010). In contrast, *igapós* drain the Archaic or Precambrian formations of the Guyana and Central Brazilian Shields.

Igapó substrates originate from erosional processes of strongly weathered and lixiviated tertiary sediments (Furch 1997). They often consist of almost pure quartz sand near the river channels, and can contain a moderate amount of silt and kaolinitic clay at densely forested, higher positions of the floodplain. Inside the forest, clayish topsoils hardly reach depths of more than 1 m. A deep litter layer and superficial root mats of trees might also develop to depths of 0.5 m (Wittmann 2012). Black-water *igapó* substrates are one of the most nutrient-poor substrates in the Amazon. With the exception of nitrogen, major nutrient concentrations in *igapó* soils, such as phosphorous, potassium, calcium and magnesium, are usually 10- or many-fold lower in *igapó* substrates than in *várzea* (Furch 1997), and thus are comparable to podzols (Coomes 1997; Targhetta *et al.* 2015) (campina and campinarana *sensu* Prance 1975). True *igapós* include regularly flood-pulsing rivers such as the Negro, Tefé and Uatumã, in the case of black waters, and the Tapajós, Trombetas, Xingu, Araguaia-Tocantins and Branco, in the case of clear waters. While most of the thousands of high-order central and western Amazonian *terra firme* creeks have similar water chemistry to *igapós* they are not defined as such. These riparian zones, called "baixios" in Brazilian Amazonia, have episodic and unpredictable water-level oscillations that respond mainly to local rainfall events (Junk 1989).

In contrast, *várzea* substrates originate from the deposition of sediment loads of the Andes and the pre-Andean foothills. The suspended sediment load is relatively fine-grained and overwhelmingly composed of clay, silt and only minor amounts of fine sand (Irion *et al.* 1997). The deposits form nutrient-rich substrates comprised of relatively unweathered clay minerals with relatively high cation exchange capacity, such as montmorillonite, smectite, illite and chlorite (Gibbs 1967; Irion 1984; Irion *et al.* 1997; Junk *et al.* 2012). While *várzea* is mostly clayey, small-scale heterogeneity is typical, induced by variation in water velocity of meandering channels that sorts sediments according to size and weight. Thus, as a result of lower water velocities, floodplain sites far from the main channels, such as scroll lakes and abandoned river meanders, are often characterized by finer substrates. In contrast, sites near the main river channels are usually characterized by slightly coarser (fine sand) substrates. Vegetation can also influence substrate by reducing local water velocities, thus trapping suspended sediment loads. This "biogenical silting up" of vegetated sites is associated with a successional sequence of different aged forests (Wittmann *et al.* 2004).

Differing nutrient conditions of *igapó* and *várzea* influence patterns of plant distribution, productivity and diversity. These differences are most visually apparent in the herbaceous plant community. In *várzea*, semi-aquatic herbaceous plants are dense and productive. They are most abundant where mean flood height and duration exceed the limit of woody species, which in *várzea* corresponds to a mean annual flood height of 7.5 m, or a duration of approximately 300 days year^{-1} (Wittmann *et al.* 2004). The herbaceous layer is often dominated by grasses, and some of these are among the most productive on Earth with net primary production of more than 100 t ha year^{-1}, as is the case for *Echinochloa polystachia* and *Paspalum repens* (Poaceae) (Piedade *et al.* 1991; 1992). In contrast, herbaceous plants in *igapó* are mainly characterized by their low density and low productivity. Slow-growing sedges (Cyperaceae), more tolerant of nutrient-poor substrates, replace the highly productive grasses of *várzea* (Junk *et al.* 2015).

Similar patterns characterize floodplain trees. *Várzea* tree species can show high population densities and exceptionally fast growth rates (Nebel *et al.* 2001; Schöngart *et al.* 2007; 2010). The rate of accumulation of organic material by vegetation, or above-ground net primary production (ANPP), can reach up to 31.8 Mg ha year^{-1} in *várzea*, more than three-fold higher than *terra firme* forest and among the highest values for tropical forest types worldwide, including plantations (Schöngart *et al.* 2010). Comparable data for *igapó* forests are lacking, however, trees generally growing much more slowly – even within species, annual tree growth rates have been shown to decrease three-fold in *igapó* relative to *várzea* (Fonseca Júnior *et al.* 2009; Schöngart *et al.* 2005).

Differences in the nutrient-status of *várzea* and *igapó* are reflected in patterns of species life history traits. For example, early-successional tree species in Amazonian *várzea*, such as *Salix martiana* (Salicaceae), *Alchornea castaneifolia* (Euphorbiaceae) or *Cecropia latiloba* (Urticaceae), have life-cycles of only 10–15 years (Schöngart 2003; Worbes *et al.* 1992). In comparable forest stages of *igapó*, early-successional tree species may reach ages of up to 60 years, such as the case for *Symmeria paniculata* (Polygonaceae), *Eugenia* spp. (Myrtaceae) and *Malouetia* spp. (Apocynaceae) (J. Schöngart, unpubl. data). Even in late-successional *várzea* forest, maximum tree ages hardly surpass 300 years (Schöngart *et al.* 2002, 2010; Worbes 1989; Worbes *et al.* 1992), while in *igapó* ages of up to 1,000 years have been documented for slow-growing *Eschweilera tenuifolia* (Lecythidaceae) (unpublished data from J. Schöngart in Junk *et al.* 2015).

Reproductive strategies are also likely to be linked to nutrient differences, especially in regards to seed size and mass. For example, many *igapó* tree species have large and heavy-weighted seeds that remain attached to seedlings for years, presumably delivering a nutrient supply and enabling seedlings to be less dependent on soil nutrients during establishment (Parolin 2000). In contrast, many *várzea* tree species have small seeds that are dispersed by wind and water (Parolin *et al.* 2013).

Differences in leaf attributes among *várzea* and *igapó* are thought to reflect strong growth-defense tradeoffs – while leaves are “cheap” and disposable in nutrient-rich *várzea*, leaves in *igapó* are relatively more costly, and thus longer lasting and better protected against herbivores (Coomes 1997; Medina *et al.* 1978; Prance 1975). Leaves in *igapó* are usually small, scleromorphic and vertically oriented, a suite of attributes shared with the flora of similarly nutrient-impoverished white-sand soils (Fine *et al.* 2004; Targhetta *et al.* 2015). Oppositely, leaves in *várzea* are mostly short-lived, deciduous during the aquatic phase, and N- and P-rich (Furch 1997) with higher photosynthetic capacities (Parolin 2009). Differences in leaf attributes scale up to affect ecosystem functioning. For example, decomposition of litter fall may take up to several years in *igapó* due the high carbon content of scleromorphic leaves, leading to a build-up of deep litter layers. In contrast, *várzea* leaves are easily decomposed and organic horizons are generally absent or not well-developed, except where permanent inundation inhibits microbial activity (e.g. in chavascal *sensu* Wittmann *et al.* 2004).

Root distribution, depth and longevity are also influenced by the contrasting nutrient conditions. In *igapó*, fine roots responsible for nutrient uptake are generally shorter, slower-growing and longer-lived (Meyer *et al.* 2010). These are mostly distributed as a thick superficial root mat, resulting in efficient capture of scarce nutrients as they enter the substrate from the litter horizon. In contrast, fine roots of *várzea* trees are faster-growing and shorter-lived (Meyer *et al.* 2010). Furthermore, they are distributed to greater depths than in *igapó*, with high densities even at 40 cm depth or more, and individual roots may reach to depths of 6 m (Wittmann and Parolin 2005).

Contrasting nutrient conditions in *várzea* and *igapó* are paralleled by differences in substrate texture, an additional factor influencing the distribution of floodplain tree species, especially with regards to drought susceptibility. Because sandier *igapó* substrates desiccate quickly, most tree seedlings are subject to intense drought during low water levels. Drought-avoiding mechanisms, such as thick cuticulas, sunken stomata, wax deposits and the development of papillae and hairs on leaf surfaces are thus especially common in *igapó* floodplains (Parolin *et al.* 2010; Waldhoff and Furch 2002; Waldhoff and Parolin 2010). *Várzea* substrates, in contrast, have comparatively high water retention capacity, which reduces desiccation even when periods of low floodwater coincide with low precipitation, as for example during El Niño years (Schöngart and Junk 2007).

6.3.4 Trees, hydro-geomorphic disturbance and light regimes

Disturbance as a result of fluvial dynamics is ubiquitous among river floodplains; however, it is especially prevalent in *várzea*. Most white-water rivers move their channels relatively quickly as a result of high sediment loads, comparatively low current velocities and meandering activity. Next to the main river channels, sedimentation on point bars may reach 30–100 cm year^{-1} (Campbell *et al.* 1992; Junk 1989). Oppositely, several hectares of forest can be eroded during a single high-water period on undercut slopes (Wittmann *et al.* 2004). These processes result in channel migration that continuously destroys and recreates floodplain areas (Kalliola *et al.* 1991). Rates of channel migration in Amazonian white-water rivers increase exponentially from the east to the west, ranging from 0.8% year^{-1} at the Solimões river near Manaus (Mertes *et al.* 1995) to 2,5 % year^{-1} at the Solimões river in western Brazilian Amazonia (Peixoto *et al.* 2009) to 14–23% along the Ucayali-Marañon in Peru (Kalliola *et al.* 1992).

Spatial and temporal change resulting from these hydro-geomorphic processes is a crucial factor influencing the *várzea* vegetation, in that it induces a natural ecosystem disturbance that maintains different successional stages (Kalliola *et al.* 1991; Peixoto *et al.* 2009; Richards *et al.* 2002; Salo *et al.* 1986; Wittmann *et al.* 2004). This is one of the most important contrasts of white-water forests to other

Amazonian floodplain and upland forests: whereas small tree-fall gaps or thunderstorm "blow downs" predominate disturbance-mediated succession in other Amazonian habitats (Chambers *et al.* 2009; Marra *et al.* 2014), hydro-geomorphic disturbance in *várzea* is more severe and occurs continuously at the landscape scale. By opening large areas for primary succession, river dynamics create a patchwork of different-aged stands that are colonized by different species, leading to exceptionally high beta diversity within Amazonian *várzea* (Campbell *et al.* 1992; Wittmann *et al.* 2006). In light of this, forest disturbance caused by modern and past fluvial dynamics is thought to increase regional biological diversity, either by reducing competitive exclusion, or even serving as a mechanism of speciation (Salo *et al.* 1986; Sedell *et al.* 1990; Wittmann *et al.* 2006).

Many *várzea* tree species are well-adapted to hydro-geomorphic dynamics, in the sense that they are fast colonizers or regenerate well after disturbance. The first *várzea* colonists of freshly deposited sediments are light-demanding and highly flood-tolerant pioneer species, such as *S. martiana* (Salicaceae), *A. castaneifolia* (Euphorbiaceae), *Tessaria integrifolia* (Asteraceae) and *C. latiloba* (Urticaceae) (Wittmann *et al.* 2004; 2010). These tree species only establish at sites where relative photosynthetic active radiation (rPAR) is 70–100% (Wittmann and Junk 2003). Attributes that make them effective colonizers include rigorous vegetative and sexual reproduction (Puhakka and Kalliola 1993; Worbes 1997), continuous production of small and wind- or water-dispersed seeds (Oliveira Wittmann *et al.* 2010; Parolin *et al.* 2013), buoyant seeds that germinate while floating (Oliveira Wittmann *et al.* 2007), the ability to produce new fine-root layers above the annual deposits, stilt roots that increase aeration during inundation (Wittmann and Parolin 2005), low wood density, a short life cycle, and fast growth (Worbes *et al.* 1992). The stems and roots of these pioneer tree species promote sedimentation by creating drag and reducing the energy of flowing water for carrying loads. As loads are deposited, the relative topographic position of vegetated stands increases (Wittmann *et al.* 2002, 2004). Following the classic model of forest succession proposed by Connell and Slatyer (1977), pioneer tree species subsequently shade their environment, inhibit the establishment of light-demanding grasses and facilitate the establishment of other, moderately light-demanding tree species. After the establishment of early-secondary tree species, pioneer shrubs no longer regenerate, because the understory light conditions are no longer suitable below 30% rPAR (Wittmann *et al.* 2010).

In *igapó*, hydro-geomorphic processes are much less intense, for both black- and clear-water rivers. In contrast to the fast-growing pioneer shrubs of *várzea*, shrubs and small trees that colonize *igapó* floodplains have comparatively slow growth rates. Thus primary succession in *igapó* may take decades. While within *várzea*, river dynamics largely drive successional processes and variation in plant composition; this is not as clearly evident in *igapó*. Indeed, *igapó* river channels can remain stable for millennia (Junk *et al.* 2015) and patterns in beta diversity may be primarily controlled by other factors or occur at other spatial scales. For example, although tree alpha diversity in *igapó* decreases with flooding severity, the impact of the flood gradient on species turnover may be relatively subtle (Ferreira 1997; Targhetta *et al.* 2015; Wittmann *et al.* 2010). Rather, variation in tree species composition is potentially more related to substrate texture (i.e. communities on clay vs. sand substrates) and local nutrient availability. Montero *et al.* (2014) argued that species composition of *igapó* forest along the Negro river occurs mainly at the scale of entire watersheds, controlled by differences in underlying geology.

6.3.5 Trees and wetland microclimates

Floodplains generate local variation in microclimate and thus can buffer adverse climatic conditions. For example, increased water availability may buffer trees from drought (Meave *et al.* 1991; Oliveira-Filho and Ratter 1995). Also, the high latent heat capacity of water buffers trees from both cold and heat (e.g. Naiman *et al.* 1993; Sculthorpe 1985). These microclimatic differences may permit wetland populations to persist in otherwise unsuitable climates. For example, a preliminary analysis of *várzea* tree species distributions shows that approximately 30% have elevation optima in cooler and wetter montane Andean forests (>1,500 m) (F. Wittmann, unpublished data). In light of the predicted hotter and drier climate in the Amazon, such results make a strong case that floodplain occupation may be an important response for drought-sensitive tree species. Indeed, drought is a primary driver of species distributions in the Amazon, leading some researchers to warn of widespread regional die-back and biotic attrition (e.g. Condit *et al.* 2013; Engelbrecht *et al.* 2005; Lewis *et al.* 2011; Parolin 2009). To what extent Amazonian floodplains might mitigate response to future climate, and for which species, has not been studied.

Because climate exerts strong controls on species distributions, the ways in which Amazonian wetlands alter local conditions also offers novel opportunity for biogeographic dispersal. Indeed, a long-standing curiosity of the floodplain plant community has been the atypical biogeographic patterns of many taxa (Steyermark 1979). For example, Andean and montane indicator genera such as *Cestrum, Clusia, Cybianthus, Ilex, Hedyosmum, Myrcia, Nectandra, Salix* and *Tessaria* are extremely rare or absent in non-flooded lowlands, but can become dominant in floodplains (Gentry 1982; Householder *et al.* 2015). Likewise, floristic parallels with dry savanna biomes have been repeatedly noted (Prance 1979; Kubitzki 1989). One way to explain the numerous taxa from other biogeographical regions is that over evolutionary time they have tracked suitable environmental conditions in floodplains that are consistent with their conserved ecological niches. For example, Prance (1979) and Kubitzki (1989) argued that taxa from savanna biomes are "pre-adapted" to seasonal flood pulses, proposing that the physiological response to seasonality in drought or flooding may be similar. Householder *et al.* (Chapter 8) further explore how biogeographic dispersal may have shaped local wetland plant communities, and even the broader Amazon region itself.

6.4 Conclusions

Amazonian floodplains are unique in their combination of:

1. extensive area;

2. strong hydro-edaphic gradients;
3. seasonal flood pulse;
4. severe disturbance regime; and
5. long-term landscape persistence.

We have argued that together these conditions play out in many ways to shape patterns of biogeography and biodiversity of floodplain vegetation and of the Amazon region itself. Across the Amazon Basin, broad-scale environmental and biological variation is underlain by a vast paleo-alluvial template. On top of this template, floodplains produce spatial and temporal heterogeneity by physically reworking the landscape and altering local environmental conditions. In so doing they maintain regional biodiversity by

a. sustaining high levels of beta diversity;
b. reducing competitive exclusion; and
c. providing microclimate refugia.

Furthermore, in light of the large number of floodplain specialists, we have explored scenarios by which Amazonian floodplains might even generate new species. Many Amazonian floodplain specialist tree species may evolve sympatrically from local *terra firme* stock as novel, flood-adapted ecotypes (Kubitzki 1989), while others have arrived "pre-adapted" to floodplains with adaptations from other biogeographic regions.

The response of the apparently narrowly-adapted floodplain flora to growing human development of the Amazon's river network is uncertain. Nearly half of the world's large rivers that once had extensive floodplains are now regulated for flood control, irrigation and hydropower generation (Lehner *et al.* 2011; Nilsson *et al.* 2005). These global trends have had dire effects on floodplain vegetation communities, because they alter natural disturbance and flood regimes (Nilsson *et al.* 2005; Pelicice *et al.* 2014). In the Amazon Basin, most rivers are currently slated for dam construction for hydropower, and plans for the construction of more than 150 dams are at advanced stages (Finer and Jenkins 2012). Growing demand for dams and their potential effects on flooding patterns may portend dire consequences for a specialist, endemic floodplain flora, especially considering that it evolved in response to the Amazon's fluvially dynamic and flood-pulsing hydrological network. In the case that loss of unique floodplain habitat leads to extinction of its specialist tree community, it is not clear how floodplain forests and their ecosystem services might be restored: no other tree species on Earth are likely to be capable of filling these niches.

References

Anderson, J., Saldana Rojas. J. and del Busto Rojas, C. (2005) Seed dispersal by a neotropical fruit-eating fish. ESA Meeting Montreal. http://abstracts.co.allenpress.com/pweb/esa2005/document/?ID=49617

Antonelli, A. and Sanmartín, I. (2011) Why are there so many plant species in the Neotropics? *Taxon*, **60**, 403–414.

Armstrong, W., Brändle, R.A. and Jackson, M.B. (1994) Mechanisms of flood tolerance in plants. *Acta Botanica Neerlandica*, **43**, 307–358.

Baker, T.R., Pennington, R.T., Magallon, S. *et al.* (2014) Fast demographic traits promote high diversification rates of Amazonian trees. *Ecology Letters*, **17**, 527–536.

Betbeder, J., Gond, V., Frappart, F. *et al.* (2014) Mapping of central Africa forested wetlands using remote sensing. *IEEE Journal of selected topics in Applied Earth Observation and Remote Sensing*, **7**, 532–542.

Burnham, R.J. and Graham, A. (1999) The history of neotropical vegetation: new developments and status. *Annals of the Missouri Botanical Garden*, **86**, 546–589.

Burnham, R.J. and Johnson, K.R. (2004) South-American palaeobotany and the origins of neotropical rainforests. *Phil. Trans. R. Soc. Lond. B*, **359**, 1595–1610.

Campbell, D.G., Stone, J.L. and Rosas, A. (1992) A comparison of the phytosociology and dynamics of three floodplain (*várzea*) forests of known ages, Rio Juruá, western Brazilian Amazon. *Botanical Journal of the Linnean Society*, **108**, 213–237.

Chambers, J.Q., Robertson, A.L., Carneiro, V.M..C. *et al.* (2009) Hyperspectral remote detection of niche partitioning among canopy trees driven by blowdown gap disturbances in the Central Amazon. *Oecologia*, **160**, 107–117.

Condit, R., Engelbrecht, B.M.J., Pino, D. *et al.* (2013) Species distributions in response to individual soil nutrients and seasonal drought across a community of tropical trees. *PNAS*, **110**, 5064–5068.

Connell, J.H. and Slatyer, R.O. (1977) Mechanisms of succession in natural communities in their role in community stability and organization. *American Naturalist*, **111**, 1119–1144.

Coomes, D.A. (1997) Nutrient status of Amazonian caatinga forests in a seasonally dry area: nutrient fluxes in litter fall and analyses of soils. *Canadian Journal of Forestry Research*, **27**, 831–839.

Correa, S.B., Winemiller, K.O., López-Fernández, H. and Galeti, M. (2007) Evolutionary perspectives on seed consumption and dispersal by fishes. *BioScience*, **57**, 748–756.

Correa, S.B., Costa-Pereira, R., Fleming, T. *et al.* (2015) Neotropical fish-fruit interactions: eco-evolutionary dynamics and conservation. *Biological Reviews*, **90**, 1263–1278.

Crawford, R.M.M. (1989) The anaerobic retreat, in *Studies in Plant Survival: Ecological case histories of plant adaptation to adversity. Studies in Ecology 11* (ed. R.M.M. Crawford), Blackwell Scientific Publ, Oxford, UK, pp. 105–129.

Crawford, R.M.M. (1992) Oxygen availability as an ecological limit to plant distribution. *Advances in Ecological Research*, **223**, 93–185.

De Simone, O., Haase, K., Müller, E. *et al.* (2002) Impact of root morphology on metabolism and oxygen distribution in roots and rhizosphere from two Central Amazon floodplain tree species. *Functional Plant Biology*, **29**, 1025–1035.

Engelbrecht, B.M.J., Kursar, T.A. and Tyree, M.T. (2005) Drought effects on seedling survival in a tropical moist forest. *Trees*, **19**, 312–321.

Evrard, C. (1968) *Recherches ecologiques sur le peuplement forestier des sols hydromorphes de la Cuvette Congolaise*, Des Presses des Ets Wellens-Pay, Paris.

Fernandez, M.D., Pieters, A., Donoso, C. *et al.* (1999) Seasonal changes in photosynthesis of trees in the flooded forest of the Mapire River. *Tree Physiology*, **19**, 79–85.

Ferreira, C.S., Piedade, M.T.F., Junk, W.J. and Parolin, P. (2007) Floodplain and upland populations of Amazonian Himatanthus sucuuba: effects of flooding on germination, seedling growth, and mortality. *Environmental and Experimental Botany*, **60**, 477–483.

Ferreira, C.S., Piedade, M.T.F., Tiné, M.A.S. *et al.* (2009) The role of carbohydrates in seed germination and seedling establishment of Himatanthus sucuuba, an Amazonian tree with populations adapted to flooded and non-flooded conditions. *Annals of Botany*, **104**, 1111–1119.

Ferreira, C.S., Figueira, A.V.O., Gribel, R. *et al.* (2010) Genetic variability, divergency and speciation in trees of periodically flooded forests of the Amazon: a case study on Himatanthus sucuuba (Spruce) Woodson, in *Amazonian Floodplain Forests. Ecophysiology, biodiversity and sustainable management. Ecological Studies 210* (eds W.J. Junk, M.T.F. Piedade, F. Wittmann *et al.*), Springer Heidelberg, London/New York, pp. 301–312.

Ferreira, L.V. (1997) Effects of the duration of flooding on species richness and floristic composition in three hectares in the Jaú National Park in floodplain forests in central Amazonia. *Biodiversity and Conservation*, **6**, 1353–1363.

Figueiredo, J., Hoorn, C., van der Veen, P. and Soares, E. (2010) Late Miocene onset of the Amazon River and the Amazon deep-sea fan: evidence from the Foz do Amazonas Basin. *Geology*, **37**, 619–622.

Fine, P.V.A., Mesones, I. and Coley, P.D. (2004) Herbivores promote habitat specialization by trees in Amazonian forests. *Science*, **305**, 663–665.

Fine, P.V.A., Zapata, F. and Daly, D.C. (2014) Investigating processes of neotropical rain forest tree diversification by examining the evolution and historical biogeography of the Protiae (Burseraceae). *Evolution*, **68**, 1988–2004.

Finer, M. and Jenkins, C.N. (2012) Proliferation of hydroelectric dams in the Andean Amazon and implications for Andes-Amazon connectivity. *PLOS ONE*, **7** e35126, 1–7.

Fonseca Júnior, S.F., Piedade, M.T.F. and Schöngart, J. (2009) Wood growth of *Tabebuia barbata* (E. Mey.) Sandwith (Bignoniaceae) and *Vatairea guianensis* Aubl. (Fabaceae) in Central Amazonian black-water (*igapó*) and white-water (*várzea*) floodplain forests. *Trees*, **23**, 127–134.

Franckland, B., Bartley, M.R. and Spence, D.H.N. (1987) Germination under the water, in *Plant Life in Aquatic and Amphibious Habitats*, Special publication series of the British Ecological Society, vol. **5** (ed. R.M.M. Crawford), Blackwell Scientific, Oxford, UK, pp. 167–178.

Furch, K. (1997) Chemistry of *várzea* and *igapó* soils and nutrient inventory of their floodplain forests, in *The Central Amazon Floodplain: Ecology of a pulsing system. Ecological Studies 126* (ed. W.J. Junk), Springer Berlin, Heidelberg/New York, pp. 47–68.

Gentry, A. (1982) Neotropical floristic diversity: phytogeographical connections between Central and South America, Pleistocene climatic fluctuations, or an accident of Andean orogeny. *Annals of the Missouri Botanical Garden*, **69**, 557–593.

Greb, F.G., DiMichele, W.A. and Gastaldo, R.A. (2006) Evolution and importance of wetlands in earth history, in *Wetlands Through Time: Geological Society of America Special Paper 399* (eds F.G. Greb and R.A. DiMichele), Geological Society of America, Boulder, CO, pp. 1–40.

Gibbs, R.J. (1967) Amazon River: environmental factors that control its dissolved and suspended load. *Science*, **156**, 1734–1737.

Gottsberger, G. (1978) Seed dispersal by fish in the inundated region of Humaitá, Amazonia. *Biotropica*, **10**, 170–183.

Hoorn, C., Wesselingh, F.P., Ter Steege, H. *et al.* (2010) Amazonia through time: Andean uplift, climate change, landscape evolution and biodiversity. *Science*, **330**, 927–931.

Householder, J.E., Wittmann, F., Tobler, M.W. and Janovec, J.P. (2015) Montane bias in lowland Amazonian peatlands: plant assembly on heterogeneous landscapes and potential significance to palynological inference. *Palaeogeography, Palaeoclimatology, Palaeoecology*, **423**, 138–148.

Irion, G. (1984) Sedimentation and sediments of Amazonian rivers and evolution of the Amazonian landscape since Pliocene times, in *The Amazon: Limnology and landscape ecology of a mighty tropical river and its basin* (ed. H. Sioli), Dr W. Junk Publishers, Dordrecht, pp. 201–214.

Irion, G. and Kalliola, R. (2010) Long-term landscape development processes in Amazonia, in *Amazonia, Landscape and Species Evolution: A look into the past* (eds C. Hoorn and F.P. Wesselingh), Blackwell Publishing, Oxford, UK, pp. 185–197.

Irion, G., Junk, W.J. and Mello, J.A.S.N. (1997) The large central Amazonian river flooplains near Manaus: geological, climatological, hydrological, and geomorphological aspects, in *Ecological Studies 126* (ed. W.J. Junk)(ed.), The Central Amazon Floodplains: Ecology of a pulsing system, Springer Verlag, Berlin/Heidelberg/New York, pp. 23–46.

Irion, G., Mello, J.A.S.N., Morais, J. *et al.* (2010) Development of the Amazon valley during the middle to late Quaternary: sedimentological and climatological observations, in *Amazonian Floodplain forest, Ecophysiology, Biodiversity and Sustainable Management. Ecological Studies 210* (eds W.J. Junk, M.T.F. Piedade, F. Wittmann *et al.*), Springer Berlin, Heidelberg/New York, pp. 27–42.

Jackson, M.B. and Colmer, T.D. (2005) Response and adaptation by plants to flooding stress. *Annals of Botany*, **96**, 501–505.

Joly, C.A. and Crawford, R.M.M. (1982) Variation in tolerance and metabolic responses to flooding in some tropical trees. *Journal of Experimental Botany*, **33**, 799–809.

Junk, W.J. (1989) Flood tolerance and tree distribution in central Amazonian floodplains, in *Tropical Forests: Botanical dynamics, speciation and diversity* (eds L.B. Holm-Nielsen, Nielsen, I.C. and H. Balslev)(eds), Academic Press, London, pp. 47–64.

Junk, W.J., Bayley, P.B. and Sparks, R.E. (1989) The flood pulse concept in river-floodplain systems. *Proceedings of the International Large River Symposium, Ottawa. Canadian Special Publications of Fisheries and Aquatic Sciences*, **106**, 110–127.

Junk, W.J., Piedade, M.T.F., Schöngart, J. *et al.* (2011) A classification of major naturally-occurring Amazonian lowland wetlands. *Wetlands*, **31**, 623–640.

Junk, W.J., Piedade, M.T.F., Schöngart, J. and Wittmann, F. (2012) A classification of major natural habitats of Amazonian white-water river floodplains (*várzea*). *Wetlands Ecology and Management*, **20**, 461–475.

Junk, W.J., Wittmann, F., Schöngart, J. and Piedade, M.T.F. (2015) A classification of the major habitats of Amazonian black-water river floodplains and a comparison with their white-water counterparts. *Wetlands Ecology and Management*, **23**, 677–693.

Kaandorp, R.J.G., Vonhof, H.B., Wesselingh, F.W. *et al.* (2005) Seasonal Amazonian rainfall variation in the Miocene Climate Optimum. *Palaeogeography, Palaeoclimatology, Palaeoecology*, **221**, 1–6.

Kalliola, R., Salo, J., Puhakka, M. and Rajasilta, M. (1991) New site formation and colonizing vegetation in primary succession on the western Amazon floodplains. *Journal of Ecology*, **79**, 877–901.

Kalliola, R., Salo, J., Puhakka, M. *et al.* (1992) Upper Amazon channel migration: implications for vegetation disturbance and succession using bitemporal Landsat MSS images. *Naturwissenschaften*, **79**, 75–79.

Kozlowski, T.T. and Pallardy, S.G. (1997) *Growth Control in Woody Plants*, Academic Press, San Diego, CA.

Kubitzki, K. (1985) The dispersal of forest plants, in (eds G.T. Prance and T.E. Lovejoy), *Amazonia – Key environments*. Pergamon, Oxford, UK, pp. 129–163.

Kubitzki, K. (1989) The ecogeographical differentiation of Amazonian inundation forests. *Plant Syst. Evol.*, **163**, 285–304.

Kubitzki, K. and Ziburski, A. (1994) Seed dispersal of floodplain forest in Amazonia. *Biotropica*, **26**, 30–43.

Lehner, B., Liermann, C.R., Revenga, C. *et al.* (2011) High-resolution mapping of the world's reservoirs and dams for sustainable river-flow management. *Frontiers in Ecology and the Environment*, **9**, 494–502.

Lewis, S.L., Brando, P.M., Phillips, O.L. *et al.* (2011) The 2010 Amazon drought. *Science*, **331**, 554.

Mannheimer, S., Bevilacqua, G., Caramaschi, E.P. and Scarano, F.R. (2003) Evidence for seed dispersal by the catfish *Auchenipterrichthys longimanus* in an Amazonian lake. *Journal of Tropical Ecology*, **19**, 215–218.

Marra, D.M., Chambers, J.Q., Higuchi, N. *et al.* (2014) Large-scale wind disturbances promote tree diversity in a Central Amazon forest. *PLOS ONE*, **9**, e103711, 1–16.

Meave, J., Kellman, M., MacDougall, D. and Rosales, J. (1991) Riparian habitats as tropical forest refugia. *Global Ecology and Biogeography Letters*, **1**, 69–76.

Medina, E., Sobrado, M. and Herrera, R. (1978) Significance of leaf orientation for leaf temperature in an Amazonian sclerophyll vegetation. *Radiation and Environmental Biophysics*, **15**, 131–140.

Melack, J.M. and Hess, L.L. (2010) Remote sensing of the distribution and extent of wetlands in the Amazon basin, in *Amazonian Floodplain Forest: Ecophysiology, biodiversity and sustainable management. Ecological Studies 210* (eds W.J. Junk, M.T.F. Piedade, F. Wittmann *et al.*), Springer Berlin, Heidelberg/New York, pp. 27–42.

Mertes, L.A.K., Daniel, L.D., Melack, J.M. *et al.* (1995) Spatial patterns of hydrology, geomorphology, and vegetation on the floodplain of the Amazon river in Brazil from a Remote Sensing Perspective. *Geomorphology*, **13**, 215–232.

Meyer, U., Junk, W.J. and Linck, C. (2010) Fine root systems and mycorrhizal associations in two central Amazonian inundation forests: *igapó* and *várzea*, in *Amazonian Floodplain Forest: Ecophysiology, biodiversity and sustainable management. Ecological Studies 210* (eds W.J. Junk, M.T.F. Piedade, F. Wittmann *et al.*), Springer Berlin, Heidelberg/New York, pp. 163–178.

Montero, J.C., Piedade, M.T.F. and Wittmann, F. (2014) Floristic variation across 600 km of inundation forests (*Igapó*) along the Negro river, central Amazonia. *Hydrobiologia*, **729**, 229–246.

Mori, S. (2001) A familia da Castanha-do-Pará: Símbolo do Rio Negro, in *Florestas do Rio Negro* (eds A.A. Oliveira and D.C. Daly), UNIP, NYBG e companhia das Letras, São Paulo, pp. 119–142.

Naiman, R.J., Décamps, H. and Pollock, M. (1993) The role of riparian corridors in maintaining regional biodiversity. *Ecological Applications*, **3**, 209–212.

Nebel, G., Dragsted, J. and Salazar Vega, A. (2001) Litter fall, biomass and net primary production in flood plain forest in the Peruvian Amazon. *Forest Ecology and Management*, **150**, 93–102.

Nilsson, C., Reidy, C.A., Dynesius, M. and Revenga, C. (2005) Fragmentation and flow regulation of the world's large river systems. *Science*, **308**, 405–408.

Oliveira-Filho, A.T. and Ratter, J.A. (1995) A study of the origin of Central Brazilian forests by the analysis of plant species distribution patterns. *Edinburgh Journal of Botany*, **52**, 141–194.

Oliveira Wittmann, A., Piedade, M.T.F., Parolin, P. and Wittmann, F. (2007) Germination in four low-*várzea* tree species of Central Amazonia. *Aquatic Botany*, **86**, 197–203.

Oliveira Wittmann, A., Lopes, A., Conserva, A.S. *et al.* (2010) Seed germination and seedling establishment in Amazonian floodplain trees, in *Amazonian Floodplain Forest: Ecophysiology, biodiversity and sustainable management. Ecological Studies 210* (eds W.J. Junk, M.T.F. Piedade, F. Wittmann *et al.*), Springer Berlin, Heidelberg/New York, pp. 259–280.

Parolin, P. (2000) Seed mass in Amazonian floodplain forests with contrasting nutrient supplies. *Journal of Tropical Ecology*, **16**, 417–428.

Parolin, P. (2009) Submerged in darkness: adaptations to prolonged submergence by woody species of the Amazonian floodplains. *Annals of Botany*, **103**, 359–376.

Parolin, P., Armbrüster, M., Wittmann, F. *et al.* (2002) A review of tree phenology in Central Amazonian floodplains. *Pesquisas Botanica*, **52**, 195–222.

Parolin, P., Lucas, C., Piedade, M.T.F. and Wittmann, F. (2010) Drought responses of flood-tolerant trees in Amazonian floodplains. *Annals of Botany*, **105**, 129–139.

Parolin, P., Wittmann, F. and Ferreira, L.V. (2013) Fruit and seed dispersal in Amazonian floodplain trees: a review. *Ecotropica*, **19**, 15–32.

Pelicice, F.M., Pompeu, P.S. and Agostinho, A.A. (2014) Large reservoirs as ecological barriers to downstream movements of Neotropical migratory fish. *Fish and Fisheries*, **16**, 697–715.

Peixoto, J.M.A., Nelson, B.W. and Wittmann, F. (2009) Spatial and temporal dynamics of alluvial geomorphology and vegetation in central Amazonian white-water floodplains by remote-sensing techniques. *Remote Sensing of Environment*, **113**, 2258–2266.

Piedade, M.T.F., Junk, W.J. and Long, S.P. (1991) The productivity of the C_4 grass *Echinochloa polystachia* on the Amazon Floodplain. *Ecology*, **72**, 1456–1463.

Piedade, M.T.F., Junk, W.J. and Mello, J..A.N. (1992) A floodplain grassland of the central Amazon, in *Primary Productivity of Grass Ecosystems of the Tropics and Subtropics* (eds S.P. Long, M.B. Jones and M.J. Roberts), Chapman & Hall, London, pp. 127–158.

Prance, G.T. (1975) Estudos sobre a vegetação das campinas Amazônicas. I: Introdução a uma série de publicações sobre a vegetação das campinas Amazônicas. *Acta Amazônica*, **5**, 207–209.

Prance, G.T. (1979) Notes on the vegetation of Amazonia. III: The terminology of Amazonian forest types subject to inundation. *Brittonia*, **3**, 26–38.

Puhakka, M. and Kalliola, R. (1993) La vegetación en áreas de inundación en la selva baja de la Amazonia Peruana, in *Amazonia Peruana: Vegetación húmeda tropical em el llano subandino* (eds R. Kalliola, M. Puhakka and W. Danjoy), Proyecto Amazonia, Turku, pp. 113–138.

Räsänen, M.E., Salo, J. and Kalliola, R. (1987) Fluvial perturbance in the western Amazon basin: regulation by long-term Sub-Andean tectonics. *Science*, **238**, 1398–1401.

Richards, K., Brasington, J. and Hughes, F. (2002) Geomorphic dynamics of floodplains: implications and a potential modeling strategy. *Freshwater Biology*, **47**, 559–579.

Rull, V. (1999) Palaeofloristics and palaeovegetacional changes across the Paleocene-Eocene boundary in northern South America. *Review of Palaeobotany and Palynology*, **107**, 83–95.

Salo, J., Kalliola, R., Häkkinen, L. *et al.* (1986) River dynamics and the diversity of the Amazon lowland forest. *Nature*, **322**, 254–258.

Scarano, F.R. (1998) A comparison of dispersal, germination and establishment of woody plants subjected to distinct flooding regimes in Brazilian flood-prone forests and estuarine vegetation, in *Ecophysiological Strategies of Xerophytic and Amphibious Plants in the Neotropics. Serie Oecologia Brasiliensis*, vol. **IV** (eds F.R. Scarano and A.C. Franco), Rio de Janeiro, PPGE-UFRJ, pp. 177–193.

Schlüter, U.-B., Furch, B. and Joly, C.A. (1993) Physiological and anatomical adaptations by joung *Astrocaryum jauari* Mart. (Arecaceae) in periodically inundated biotopes of Central Amazonia. *Biotropica*, **25**, 384–396.

Schöngart, J. (2003) *Dendrochronologische Untersuchungen in Überschwemmungswäldern der várzea Zentralamazoniens. Göttinger Beiträge zur Land- und Fortswirtschaft in den Tropen und Subtropen 149*, Erich Goltze Verlag, Göttingen.

Schöngart, J. and Junk, W.J. (2007) Forecasting the flood-pulse in Central Amazonia by ENSO-indices. *Journal of Hydrology*, **335**, 124–132.

Schöngart, J., Piedade, M.T.F., Ludwigshausen, S. *et al.* (2002) Phenology and stem-growth periodicity of tree species in Amazonian floodplain forests. *Journal of Tropical Ecology*, **18**, 581–597.

Schöngart, J., Piedade, M.T.F., Wittmann, F. *et al.* (2005) Wood growth patterns of *Macrolobium acaciifolium* (Benth.) Benth. (Fabaceae) in Amazonian black-water and white-water floodplain forests. *Oecologia*, **145**, 454–461.

Schöngart, J., Wittmann, F., Worbes, M. *et al.* (2007) Management criteria for *Ficus insipida* Willd. (Moraceae) in Amazonian white-water floodplain forests defined by tree-ring analysis. *Annals of Forest Science*, **64**, 657–664.

Schöngart, J., Wittmann, F. and Worbes, M. (2010) Biomass and net primary production of Central Amazonian floodplain forests, in *Amazonian Floodplain Forest: Ecophysiology, biodiversity and sustainable management. Ecological Studies 210* (eds W.J. Junk, M.T.F. Piedade, F. Wittmann *et al.*), Springer Berlin, Heidelberg/New York, pp. 347–388.

Sculthorpe, C.D. (1985) *The Biology of Aquatic Vascular Plants*, Koeltz Scientific Books, Königstein, Germany.

Sedell, J.R., Reeves, G.H., Hower, F.R. *et al.* (1990) Role of refugia in recovery from disturbances: modern fragmented and disconnected river systems. *Environmental Management*, **14**, 711–724.

Silva da, J.M.C., Rylands, A.B. and Fonseca da, G.A.B. (2005) The fate of the Amazonian areas of endemism. *Conservation Biology*, **19**, 689–694.

Sioli, H. (1954) Beiträge zur regionalen Limnologie des Amazonasgebietes. *Archiv für Hydrobiologie*, **45**, 267–283.

Steyermark, J.A. (1979) Plant refuge and dispersal centres in Venezuela: their relict and endemic element, in *Tropical Botany* (eds K. Larsen and L.B. Holm-Nielsen), Academic Press, London, pp. 185–221.

Targhetta, N., Kesselmeier, J. and Wittmann, F. (2015) Effects of the hydroedaphic gradient on tree species composition and aboveground wood biomass of oligotrophic forest ecosystems in the central Amazon basin. *Folia Geobotanica*, **50**, 185–205.

Terborgh, J. and Andresen, E. (1998) The composition of Amazonian forests: patterns at local and regional scales. *Journal of Tropical Ecology*, **14**, 645–664.

Visser, E..J.W., Voesenek, L.A.C.J., Vatapetian, B.B. and Jackson, M.B. (2003) Flooding and plant growth. *Annals of Botany*, **91**, 107–109.

Waldhoff, D. and Furch, B. (2002) Leaf morphology and anatomy in eleven tree species from Central Amazonian floodplains (Brazil). *Amazoniana*, **17**, 79–94.

Waldhoff, D. and Parolin, P. (2010) Morphology and anatomy of leavesIn, in *Amazonian Floodplain Forests: Ecophysiology, Biodiversity and sustainable management. Ecological Studies 210* (eds W.J. Junk, M.T.F. Piedade, F. Wittmann *et al.*)(eds), Springer Verlag, Heidelberg/Berlin/New York pp, pp. 179–202.

Waldhoff, D., Saint-Paul, U. and Furch, B. (1996) Value of fruits and seeds from the floodplain forests of Central Amazonia as food resource for fish. *Ecotropica*, **2**, 143–156.

Wittmann, F. (2012) Tree species composition and diversity in Brazilian freshwater floodplains, in *Mycorrhiza: Occurrence in natural and restored environments* (ed. M.C. Pagano), Nova Science Publ, New York, pp. 223–263.

Wittmann, F. and Junk, W.J. (2003) Sapling communities in Amazonian white-water forests. *Journal of Biogeography*, **30**, 1577–1588.

Wittmann, F. and Parolin, P. (1999) Phenology of six tree species from central Amazonian *várzea*. *Ecotropica*, **5**, 51–57.

Wittmann, F. and Parolin, P. (2005) Above-ground roots in Amazonian floodplain trees. *Biotropica*, **37**, 609–619.

Wittmann, F., Anhuf, D. and Junk, W.J. (2002) Tree species distribution and community structure of Central Amazonian *várzea* forests by remote-sensing techniques. *Journal of Tropical Ecology*, **18**, 805–820.

Wittmann, F., Junk, W.J. and Piedade, M.T.F. (2004) The *várzea* forests in Amazonia: flooding and the highly dynamic geomorphology interact with natural forest succession. *Forest Ecology and Management*, **196**, 199–212.

Wittmann, F., Schöngart, J., Montero, J.C. *et al.* (2006) Tree species composition and diversity gradients in white-water forests across the Amazon Basin. *Journal of Biogeography*, **33**, 1334–1347.

Wittmann, F., Schöngart, J. and Junk, W.J. (2010) Phytogeography, species diversity, community structure and dynamics of Amazonian floodplain forests, in *Amazonian Floodplain Forests: Ecophysiology, biodiversity and sustainable management. Ecological Studies 210* (eds W.J. Junk, M.T.F. Piedadem, F. Wittmann *et al.*), Springer Verlag, Heidelberg/Berlin/New York, pp. 61–104.

Wittmann, F., Householder, E., Piedade, M.T.F. *et al.* (2013) Habitat specifity, endemism and the neotropical distribution of Amazonian white-water floodplain trees. *Ecography*, **36**, 690–707.

Worbes, M. (1986) *Lebensbedingungen und Holzwachstum in zentralamazonischen Überschwemmungswäldern. Scripta Geobotanica 17:1–112*, Erich Goltze Verlag, Göttingen.

Worbes, M. (1989) Growth rings, increment and age of trees in inundation forests, savannas and a mountain forest in the neotropics. *IAWA Bull*, **10**, 109–122.

Worbes, M. (1997) The forest ecosystem of the floodplains, in *The Central Amazon Floodplains* (ed. W.J. Junk), Springer Berlin, Heidelberg/New York, pp. 223–266.

Worbes, M., Klinge, H., Revilla, J.D. and Martius, C. (1992) On the dynamics, floristic subdivision and geographical distribution of *varzea* forests in Central Amazonia. *Journal of Vegetation Science*, **3**, 553–564.

Ziburski, A. (1991) Dissemination, Keimung und Etablierung einiger baumarten der Überschwemmungswälder Amazoniens, in *Tropische und subtropische Pflanzenwelt*, vol. **77** (ed. W. Rauh), Akademie der Wissenschaften und der Literatur, pp. 1–96.

7

A Diversity of Biogeographies in an Extreme Amazonian Wetland Habitat

Ethan Householder, John Janovec, Mathias W. Tobler and Florian Wittmann

Abstract

Amazonian wetlands are subject to prolonged waterlogging, which is known to be a major determinant of local composition and species distributions. However, our understanding of which traits determine species distributions in Amazonian wetlands and how these traits evolved is still scant. Due to correspondence between species traits, the niche and biogeographic pattern, novel perspectives on wetland assembly processes might be gained by examining local assemblies within a biogeographic framework. Here, we consider a biogeographic framework to examine the response of woody plant assemblages to permanent waterlogging in seven lowland wetlands of peat substrate. We quantitatively examine the variability of biogeographic relations among co-existing species to show that many taxa in extreme wetlands are uncommon in surrounding forests, but rather have wider biogeographic affinities. We propose that a high diversity of biogeographic histories may provide ecologically differentiated taxa for the colonization of environmentally extreme habitats in Amazonia.

Keywords *Amazonian wetlands; biogeographic structure; hydro-edaphic gradients; hydrological regime; plant communities; soil waterlogging; species diversity; terra firme forests*

7.1 Introduction

Amazonian wetlands are associated with lower species diversity relative to surrounding *terra firme* forests, as well as compositional turnover along strong hydro-edaphic gradients (Junk *et al.* 2011). These patterns are traditionally ascribed to the edaphic conditions that filter species on the basis of their tolerance to waterlogging. Extended periods of waterlogging alter the availability of nutrients and soil oxygen to plants, as well as the type and concentration of phytotoxins (Crawford 1992). These multiple abiotic stressors associated with waterlogging hinder the growth and development of the majority of vascular plant species that occur in wetland habitats (Armstrong *et al.* 1994). Furthermore, because species differ in their ecophysiological response to soil

Forest structure, function and dynamics in Western Amazonia, First Edition.
Edited by Randall W. Myster.

waterlogging, hydrological regime is likely a major determinant of the local diversity, species distribution and assemblage of plant communities in wetland habitats (Wittmann *et al.* 2006).

Many of the phenotypic and eco-physiological features, or traits, of species that commonly inhabit severely waterlogged sites are often purported to be adaptations evolved in response to flooded conditions (Jackson and Colmer 2005; Parolin *et al.* 2000, 2004). However, an understanding of which traits, how they evolved, and how they determine species distributions along the complex hydro-edaphic gradients of Amazonian wetlands remains elusive (Pitman *et al.* 2014; Wittmann *et al.* 2013). A more profitable approach might be to examine the broader distributions of the species, genera and families of plants that inhabit wetlands, and ask whether they may have moved into wetlands from similarly extreme environments (and thus have been pre-adapted to wetlands to some degree), or are more likely to have come from the more benign *terra firme* forests (and therefore more likely to have evolved specific adaptations to wetlands) (Table 7.1). Because of an expected correspondence between species traits, the ecological niche and biogeographic pattern, the examination of local assemblies within a biogeographic framework can lead to novel insights regarding the evolutionary and historical processes that influence local assembly (Ackerly 2004; Harrison and Grace 2007; Householder 2015).

In this chapter we use a biogeographic framework to examine the response of woody plant assemblages to near-permanent waterlogging in seven lowland wetlands of peat substrates located in southern Peru. Reports of many atypical biogeographic patterns of taxa associated with these and other Amazonian wetland habitats are scattered throughout the literature, making them intriguing testing grounds to examine biogeographic structure along local environmental gradients (Gentry 1982; Householder *et al.* 2015; Kubitzki 1989; Marchant *et al.* 2002; Oliveira Filho and Ratter 1995; Prance 1979; Steyermark 1979).

Table 7.1 Loadings of the 8 physiognomic variables recorded in sample wetlands on the primary PCA axis. This axis describes a transition from taller forested sites with reduced herbaceous layers (shady understory) to shrubby sites with increasing density of smaller-stemmed, low-statured individuals and a well-developed herbaceous layer dominated by light-loving Cyperaceae and Poaceae.

Physiognomic Variable	Loadings on first PCA axis
Shrub Count	0.448
Herb Cover	0.29
Tree Count	0.273
Shrub Cover	0.262
Herb Height	0.262
Shrub Height	−0.212
Tree Cover	−0.48
Tree Height	−0.504

We first examine the variability of biogeographic relations among locally co-existing species in the sampled wetlands to show that, consistent with reports, many taxa in extreme wetland environments (and presumably their associated traits) are not common in surrounding forests, but rather have wider biogeographic origins or affinities. We hypothesize that a high diversity of biogeographic histories may provide ecologically differentiated taxa for the colonization of environmentally extreme habitat types in the Amazon. We then test predictions based on this hypothesis by examining biogeographic structure along a local physiognomic gradient within the sampled wetlands, with the expectation that more open-canopied, stunted and stressful sites are more biogeographically variable than those of more benign forested sites.

7.2 Methods

7.2.1 Habitat description

The wetlands examined in the study are vegetated peatlands located in the southern Peruvian Amazon. Peats in tropical lowlands can form in sites exhibiting perennially waterlogged substrates and minimal water level oscillations. Organic substrates derived from the biomass of plants growing *in situ* accumulate under hypoxic conditions induced by inundation (Householder *et al.* 2012). No extensive mapping of modern Amazonian wetlands on peat substrate is currently available, but Ruokolainen *et al.* (2001) provide an initial basin-wide estimate of 150,000 km^2 deduced from satellite imagery. The majority of these are expected to occupy the subsiding foreland basin region in the western Amazon, although small patches exist throughout the basin (Draper *et. al.* 2014). Habitat classifications of peatland in the Amazon are often based on the dominant plant growth form, ranging from forested habitats dominated by trees to habitats dominated by shrubs or herbs in the most extreme sites (Householder *et al.* 2012; Kalliola *et al.* 1991).

7.2.2 Vegetation sampling

We sampled seven wetlands on peat substrate along a 250 km lowland stretch of the Madre de Dios river in southeast Peru (Figure 7.1). Accumulation of thick peat (3–9 m in depth) and dominance of a single palm indicator species (*Mauritia flexuosa*) suggested relatively similar hydro-edaphic conditions characterized by perennial substrate saturation and limited water-level oscillations. Strong physiognomic transitions from swamp forest to low-statured herbaceous savannahs are readily detectable in satellite imagery and characterized all sampled wetlands (Hamilton *et al.* 2005; Householder *et al.* 2012). We systematically sampled the woody vegetation along this physiognomic gradient using nested tree and shrub plots. Trees (dbh > 10 cm) were sampled in 10 m × 10 m square plots. Woody vegetation with dbh of more than 10 cm and height of more than 1 m (shrubs and small trees) was sampled in nested 5 m × 5 m "shrub" plots positioned at the center of its companion tree plot. Total plot abundance of each species

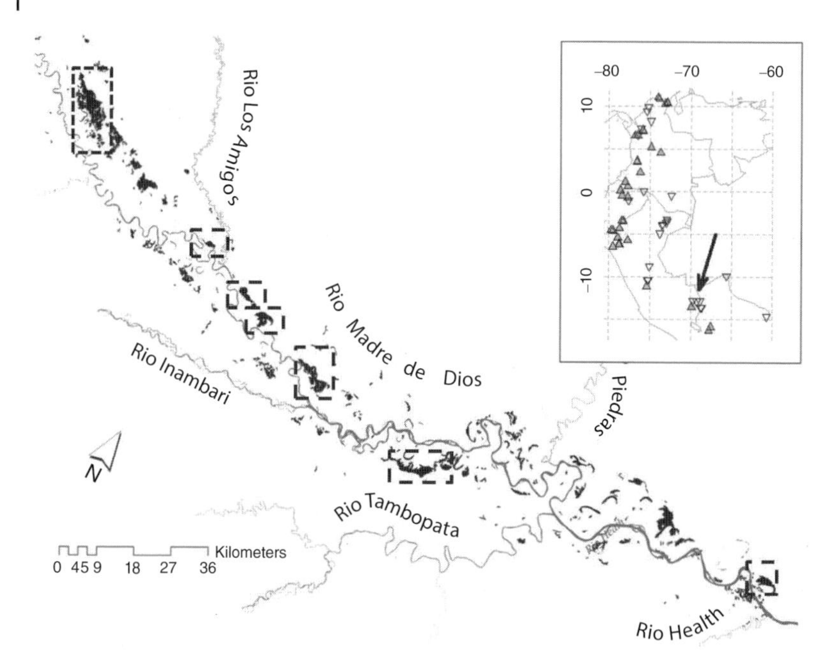

Figure 7.1 The distribution of sampled wetlands along a 250-km stretch of the Madre de Dios river, Peru. The inset shows the locations of the sampled wetlands in relation to the distribution of the Gentry plot network used to assess distributional patterns of families and genera along the elevation gradient. The Gentry sites are classified into two groups differentiated by site elevation, where upwards and downwards pointing triangles correspond to sites at greater than and less than 1,000 m a.s.l., respectively.

was estimated by summing the number of stems in both tree and nested shrub plots. Consecutive plots were placed at approximate 100 m intervals following straight-line transects, in an attempt to capture the entire physiognomic gradient at each site. Botanical collections of all species and corresponding images and metadata are made available through the Digital Herbarium of the Atrium website (http://atrium.andesamazon.org) in the "Aguajal Project".

Six structural variables, including percent coverage and height of the tree, shrub and herb strata were visually estimated in each plot. Variation in physiognomic and environmental conditions within plots was minimal. Stem density of both trees (dbh $>$ 10 cm) and shrubs (dbh $<$ 10cm) were tallied and included as additional structural variables. In total, we surveyed 147 nested plots more or less equally distributed in the seven peatlands.

For comparison with our sampled wetland sites, vegetation data was extracted from an extensive vegetation plot network compiled by Alwyn Gentry and maintained by the Missouri Botanical Garden (Phillips and Miller 2002). The Gentry transect data are based on a standard sampling method used across an extensive geographic area. Because they are representative samples of the predominant forest types, they have been used in a number of plant macroecological studies (e.g. Enquist *et al.* 2002; Punyasena *et al.* 2008). Specifically, Gentry transect data were collected in 0.1-ha plots consisting of 10 contiguous 50 m × 2 m linear subplots, where all woody individuals of less than 2.5 cm dbh were reported. Sequential subplots placed end to end were oriented in random directions. The accuracy of species determinations provided in the dataset was assessed by cross-referencing voucher numbers provided in the original dataset, with current determinations of collections stored at the Missouri Botanical Gardens and made accessible through the Tropicos website (http://www.tropicos.org/Home.aspx). Taxonomic nomenclature was standardized to APG III (2009).

Six Gentry sites from the Peruvian Amazon were chosen for comparison with the sampled wetland sites. These included three lowland (<500 m a.s.l.) sites from each of the southern and northern Peruvian Amazon. The remaining 62 Gentry sites located within the Andes–Amazon region were used to develop a biogeographic framework for examining local assembly structure.

7.3 Construction of a biogeographic framework

Floristic turnover along the Andes–Amazon elevation gradient represents one of the most outstanding features of Neotropical biogeography (Antonelli and Sanmartín 2011; Gentry 1982; Hoorn *et al.* 2010). Previous assessments of the Gentry transect data have demonstrated strong taxonomic structure along the Andes–Amazon elevation gradient, presumably driven by strong abiotic gradients that sort taxa according to highly conserved ecological niches (Enquist *et al.* 2002; Punyasena *et al.* 2008). Using the Gentry transect data, including 62 sites ranging from 0 to 3,200 m a.s.l. distributed in Colombia, Venezuela, Ecuador, Peru, Brazil and Bolivia (excluding sites located in semi-deciduous, dry forest or savanna biomes), we calculated the abundance-weighted mean elevation of each family and genus in the Gentry transect data. These weighted mean family and genus elevation values derived from the Gentry transect data were used as a quantitative measure of species' biogeographic dissemblance in local assemblies, where individuals of the same family or genus were assigned the same value reflecting their shared ancestry and biogeographic legacy. The variability of biogeographic relations among species within local assemblages was estimated using Rao's quadratic entropy (FD_Q) (Botta-Dukát 2005). In our application of it, FD_Q expresses the average difference in biogeographic pattern between two randomly selected individuals, with replacement. Thus, in cases where locally co-existing species from lineages with similar elevational distributions occur together, sites will have lower FD_Q values than those comprised of species from lineages with highly divergent elevational distributions.

We first examined differences in FD_Q between the seven sampled wetlands and six forested sites using the subset of Gentry transects that were excluded from the calculations of abundance-weighted mean elevations. The difference in mean FD_Q between the two habitats was assessed using two separate t-tests with individuals assigned either family-level or genus-level elevational means. Using the wetlands dataset in isolation, we then examined variation in FD_Q among plots distributed along the sampled physiognomic gradient. The eight physiognomic variables were first reduced to the primary axes of a principal component analysis

(PCA) of their correlation matrix. To test for association between FD_Q and the main physiognomic gradient, we used ordinary least squares regression on the first PCA axis. To ensure estimates of FD_Q were based on an adequate number of species, wetland plots with fewer than five species were removed prior to analysis. To identify individual families and genera strongly influencing FD_Q, for each wetland plot we calculated percent change in FD_Q when each taxa was left out. All analyses were done with individuals assigned to either genus-level or family-level elevational means as derived from the Gentry transect sites. Analyses were performed in R version 3.1.1 using the FD package (Lalitberté and Legendre 2010).

7.4 Results

In the sampled wetlands, we recorded 6,868 individual woody stems distributed among 165 genera and 83 families. This compares to 2,332 stems distributed among 306 genera and 88 families in the subset of 6 Gentry transects in lowland *terra firme* forests. Ranges in mean elevational distribution were similar in the sampled wetlands and the Gentry *terra firme* sites. For example, in our sampled wetlands, mean elevational distribution ranged from 100 to 2,613 m a.s.l. for families and 100 to 2,623 m a.s.l. for genera compared to 166 to 2,575 m a.s.l. and 112 to 2,407 m a.s.l. for families and genera in the *terra firme* transects. In contrast, the proportion of taxa with relatively high mean elevational distributions was greater in wetlands. For example, 50% of families and 38% of genera in the wetlands had mean elevational distributions above 1,000 m a.s.l., compared to 35% and 21% of families and genera in the lowland *terra firme* sites. There was strong evidence of a difference in mean FD_Q values between wetlands and lowland *terra firme* sites when individuals were assigned mean elevations at either family level (two-sided p value < 0.001) or genus level (two-sided p values $= 0.01$, Figure 7.2).

Within the wetlands dataset, FD_Q was associated with vegetation physiognomy. The first principal components axis of the physiognomic variables explained 28% of the variation in physiognomic variables and described the dominant physiognomic transition that we observed in the field, ranging from tall forested sites with high tree cover, shady understory and low densities of understory shrubs to stunted, high density shrubby vegetation with abundant herb cover (Figure 7.1). We interpreted this axis as a measure of relative "shrubbiness" of the vegetation. FD_Q increased with increasing shrubbiness, although in ordinary least squares regression this association was stronger at the family level ($r^2 = 0.32$, $p < 0.0001$) compared to genus level ($r^2 = 0.18$, $p < 0.0001$) (Figure 7.3a). Families and genera that had the greatest relative contribution to FD_Q were mostly mid- to high-elevation taxa of small trees and shrubs (Figure 7.3b).

7.5 Discussion

The wetland habitats we sampled – representing an extreme abiotic habitat in the Amazon – are strongly characterized by a wide variation in the biogeographic

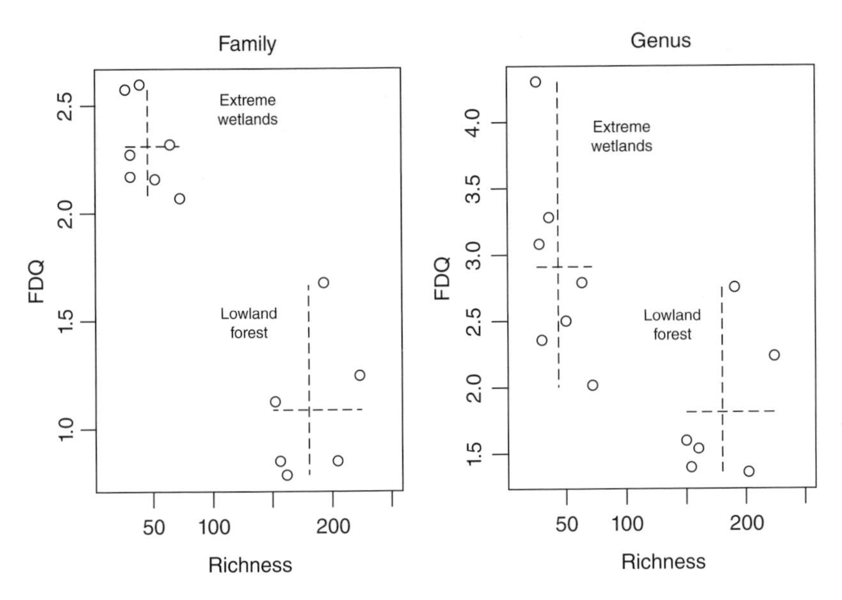

Figure 7.2 A comparison of the variability of biogeographic relations among co-occurring individuals (FD_Q) between the sampled wetland sites and 6 Gentry sites representative of lowland (<500 m a.s.l.) tropical forests. Sites are distributed along a horizontal axis of richness based on a random sub-sample of 365 individual stems (the least number of stems at a single site). Dashed horizontal and vertical lines indicate group means along both axes. While extreme wetland assemblies tend to be less species rich, their constituent taxa are highly variable in regards to their biogeographic patterns.

distribution of their constituent taxa. This variation stands in stark contrast to that of the *terra firme* – the tall, closed-canopy forested conditions on mineral substrates that overwhelmingly predominate in Amazonia. Furthermore, analyses within the sampled wetland habitat revealed that the diversity of biogeographic relations among species is not uniform along the principal environmental gradient. Rather, more biogeographically diverse plant assemblies are more commonly encountered in low-statured, and presumably more stressful, low-productivity sites than in more forested conditions. Increasing biogeographic variability was primarily a consequence of the large numbers of mid- to high-elevation taxa of small trees and shrubs. In addition to the predominantly mid-high elevation taxa identified by our sensitivity analysis (Figure 7.3b) – Araliaceae, *Cyathea* (Cyatheaceae), *Cybianthus* (Primulaceae), *Graffenrieda* (Melastomataceae), *Hedyosmum* (Chloranthaceae), *Ilex* (Aquifoliaceae), *Miconia* (Melastomataceae), Onagraceae, *Prunus* (Rosaceae), *Psychotria* (Rubiaceae) and *Schoenobiblus* (Thymeleaceae) – our plant collections indicate that many other mid-high elevation taxa are also common in the sampled wetlands, such as *Begonia* (Begoniaceae), *Centropogon* (Campanulaceae), *Cestrum* (Solanaceae), *Isoetes* (Isoetaceae), *Myrcia* (Myrtaceae), *Myrsine* (Primulaceae), *Nectandra* (Lauraceae), *Polypodium* (Polypodiaceae) and *Talauma* (Magnoliaceae).

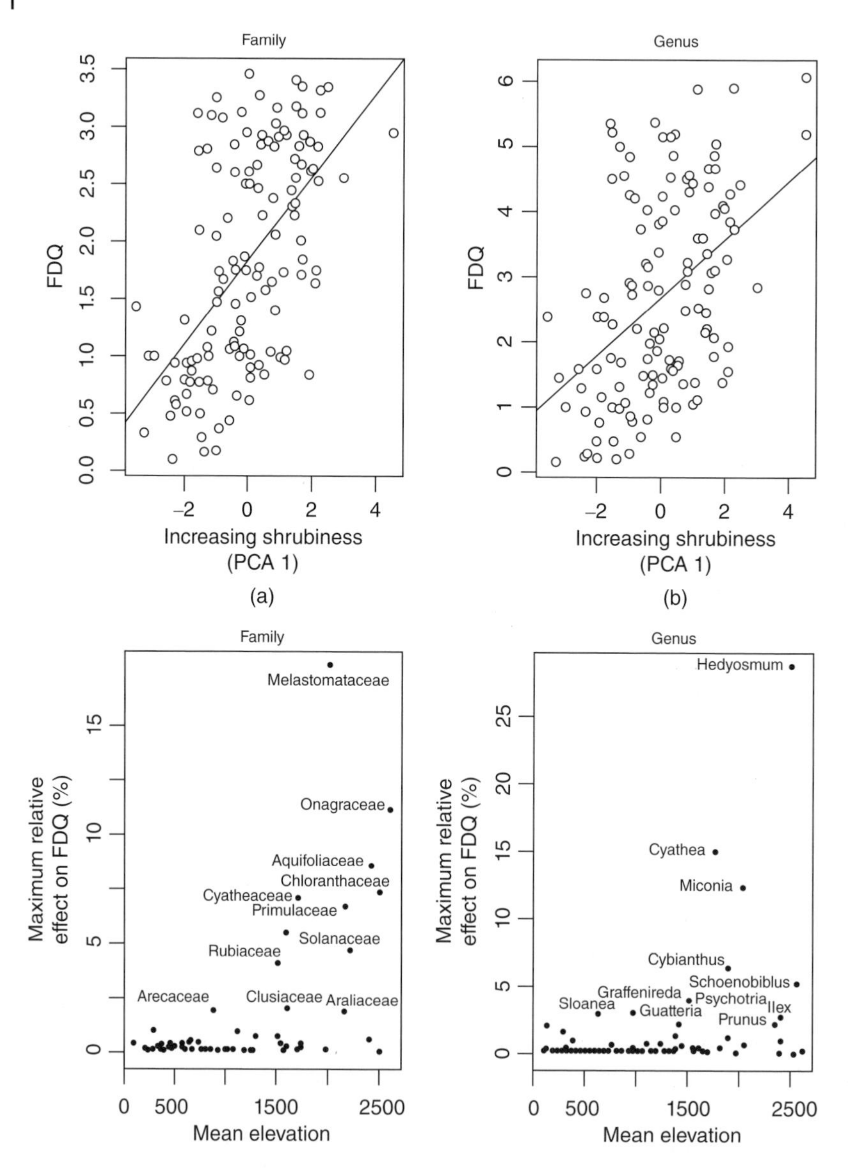

Figure 7.3 Scatterplots showing the change in biogeographic variability (FD_Q) in local plant assemblies along the main physiognomic gradient sampled in wetlands (a) and the taxa with the highest relative influence on our measure of biogeographic variability (b).

7.5.1 Insights into local assemblies

In the sense that mid- to high-elevation taxa occur more frequently and abundantly in shrubbier, more low-productivity sites, our biogeographic metric of family- and genus-level elevational distribution provided a broad predictor of species' ecological behavior along a lowland hydro-edaphic gradient. These results are consistent with the notion that biogeographic patterns result from the tendency for species to retain their ancestral niche that, in turn, determine the ability of species to colonize different environments (Ackerly 2003). We hypothesize that the occupation of the stressful, environmentally extreme ranges of hydro-edaphic gradients represents an increasingly large departure from the range of ecological strategies that have generally been successful throughout the evolutionary history of taxa within the predominant environmental condition in the Amazon – that of tall, closed-canopy forest on mineral substrate. In this regard, a high diversity of biogeographic histories may provide ecologically differentiated taxa for the colonization of environmentally extreme habitat types in the Amazon. We thus interpret the patterns found here to be a consequence of evolutionary stasis and habitat tracking of an ancestral ecological niche on a heterogeneous, lowland Amazonian landscape (e.g. Pillon *et al.* 2010).

A body of literature examining individual taxa has tended to highlight strong niche differentiation of species in edaphically specialized Amazonian habitat types and thus underscore the strong potential for underlying environmental variability to lead to phenotypic diversification and speciation (Gentry 1981; 1988; Tuomisto *et al.* 1998). Our interpretations provide different, but complementary insights into the evolution of plant communities in extreme Amazonian habitats. While the capacity for natural selection to lead to adaptation across strong environmental gradients has certainly shaped the modern vegetation composition of edaphically distinct Amazonian habitat types (Fine *et al.* 2005), our biogeographic analyses suggests that some of the unique floristic elements of these environments may be due to tracking of more ancient niche conditions rather than adaptation to novel ones. For example, there are many edaphic parallels between high-elevation forest habitat and the extreme wetlands sampled here, including low mineral nutrition, phenolic soil toxicity, reduced decomposition, low nitrogen mineralization and waterlogged soils (Bruijnzeel and Veneklaas 1998; Tanner *et al.* 1998). In similarly stressed conditions, convergent suites of attributes are found across a range of climates and ecosystems, including low rates of growth, photosynthesis, nutrient absorption and high concentrations of secondary metabolites (Coomes and Grubb 1996; 1998; Reich *et al.* 1997). Similar ecological filters between highland habitat and extreme lowland habitats are likely to impose convergent effects on species traits and to have selected for several common ecological attributes, potentially allowing species to migrate much more easily between Andean forests and Amazonian wetlands than within Amazonia itself.

7.5.2 Insights into biogeographic processes

Exchange between Andean and Amazonian floras is well known to have contributed immensely to Neotropical floristic diversity, but different plant clades have not partaken in this exchange equally (Gentry 1982). How might our results be relevant to understanding this process of biotic exchange? One potential biogeographic explanation of the patterns found in this study is that strong ecological filters in extreme wetland habitats have operated over evolutionary time to alter the regional pool from where species (and their traits) can be recruited. In such a scenario, the unique character of more extreme sites regarding their larger proportion of mid- to high-elevation taxa (and thus high biogeographic diversity) could be explained by evolutionary conservatism and adaptation from montane habitats in the Andes.

An alternative explanation is that ancient ecological differentiation of taxa on an environmentally heterogeneous lowland landscape preset the stage for subsequent biogeographic expansion. One of the major findings of this study is that the ecological differentiation of taxa with high Andean or low Amazonian biogeographic tendencies is broadly expressed along local environmental gradients in the lowlands. Thus, the major niche differences responsible for the geographic separation of Andean and Amazonian floristic groups (*sensu* Gentry 1982) on the modern geophysical template were likely to have already been in place, even prior to Andean mountain building. It is plausible that long-standing ecological filters in the lowlands could have set the stage for a selective dispersal into the newly forming Andean habitat, as new high-elevation species would have arisen from low-elevation ancestors (Gentry 1982). To the extent that this scenario is true, the extremely depauperate nature of edaphically stressful habitats in the modern lowland Amazon landscape – such as the wetlands sampled here – may not be commensurate with their historical contributions to Neotropical floristic diversity.

7.5.3 Limits of the data

Our analysis is clearly limited by the specific taxonomic and geographic windows we have applied to describe biogeographic pattern. Our framework rests on the assumption that the taxonomic groups examined here approximate natural phylogenetic and morphological divisions that are highly conserved for long periods of evolutionary time (Sepkoski and Kendrick 1993). Likewise, our examination of biogeographic pattern, limited to the Andes–Amazon region, does not represent the entire spatial distributions of all families and genera considered here. We thus assume that differences in spatial distributions along a strong elevational gradient are sufficient to broadly capture important niche differences among taxa. It is not clear how consideration of other temporal and geographic scales, or other extreme Amazonian habitat types, would alter or reinforce observed patterns. However, in other extreme Amazonian habitat types, several authors have noted increasing abundances of taxonomic groups more typical of extra-Amazonian regions – especially from dry forest and savanna habitat (Kubitzki 1989; Prance 1979; Wittmann *et al.* 2013) – suggesting that expansion of the geographic window could reveal similar pattern as that observed here.

Acknowledgements

The manuscript was developed and written with funding provided to the corresponding author by the Conselho Nacional de Desenvolvimento Científico e Tecnológico (CNPQ-process number 141727/2011-0). We thank the Max Planck Institute for Chemistry and Maria Teresa Fernandez Piedade for providing a productive, hospitable work environment. Field research was made possible by funding from the Discovery Fund of Fort Worth, Texas, the Gordon and Betty Moore Foundation (grant no. 484), the US National Science Foundation (grant no. 0717453), and the Programa de Ciencia y Tecnologia–FINCYT (co-financed by the Banco Internacional de Desarollo, BID) grant number PIBAP-2007–005. We thank Sy Sohmer, Cleve Lancaster, Pat Harrison, Will McClatchey, the board of directors, administration, development and general staff of the Botanical Research Institute of Texas (BRIT) for providing important institutional support and infrastructure over the years. We are grateful to Jason Wells, Amanda Neill, Keri Barfield and Renan Valega for logistical support, assistance and discussion during various phases of this research. We also thank Fernando Cornejo, Piher Maceda, Javier Huinga, Angel Balarezo and Fausto Espinoza for assisting in different phases of field research. We appreciate various institutions for ongoing research, collection and export permits during the course of this project, including the former Instituto Nacional de Recursos Naturales (INRENA), the Dirección General de Flora y Fauna Silvestre (DGFFS), the Ministerio de Agricultura (MINAG) and the Ministerio del Ambiente (MINAM). Finally, we thank several individuals who made helpful comments on previous versions of the manuscript, including Bill Magnusson, Camila Ribas, Hans ter Steege, Nigel Pitman, Paul Fine and Simon Queenborough.

References

Ackerly, D. (2003) Community assembly, niche conservatism, and adaptive evolution in changing environments. *International Journal of Plant Sciences*, **164** (S3), 164–184.

Ackerly, D. (2004) Adaptation, niche conservatism, and convergence: comparative studies of leaf evolution in the California chaparral. *The American Naturalist*, **163** (5), 654–671.

The Angiosperm Phylogeny Group (2009) An update of the Angiosperm Phylogeny Group classification for the orders and families of flowering plants: APG III. *Botanical Journal of the Linnean Society*, **161**, 105–121.

Antonelli, A. and Sanmartín, I. (2011) Why are there so many plant species in the Neotropics? *Taxon*, **60**, 403–414.

Armstrong, W., Brandle, R. and Jackson, M.B. (1994) Mechanisms of flood tolerance in plants. *Acta Botanica Neerlandica*, **43**, 307–358.

Botta-Dukát, Z. (2005) Rao's quadratic entropy as a measure of functional diversity based on multiple traits. *Journal of Vegetation Science*, **16**, 533–540.

Bruijnzeel, L.A. and Veneklaas, E.J. (1998) Climatic conditions and tropical montane forest productivity: the fog has not yet lifted. *Ecology*, **79**, 3–9.

Coomes, D.A. and Grubb, P.J. (1996) Amazonian Caatinga and related communities at La Esmeralda, Venezuela: Forest structure, physiognomy and floristics, and control by soil factors. *Vegetatio*, **122**, 167–91.

Coomes, D.A. and Grubb, P.J. (1998) A comparison of 12 tree species of Amazonian Caatinga using growth rates in gaps and understory, and allometric relationships. *Functional Ecology*, **12**, 426–35.

Crawford, R. (1992) Oxygen availability as an ecological limit to plant distribution. *Advances in Ecological Restoration*, **223**, 93–185.

Draper, F.C., Roucoux, K.H., Lawson, I.T. *et al.* (2014) The distribution and amount of carbon in the largest peatland complex in Amazonia. *Environmental Research Letters*, **9** (12), 124017.

Enquist, B.J., Haskell, J.P. and Tiffney, B.H. (2002) General patterns of taxonomic and biomass partitioning in extant and fossil plant communities. *Nature*, **419**, 1–4.

Fine, P., Daly, D.C., Villa Muñoz, G. *et al.* (2005) The contribution of edaphic heterogeneity to the evolution and diversity of Burseraceae trees in the Western Amazon. *Evolution*, **59** (7), 1464–1478.

Gentry, A.H. (1981) Distributional patterns and an additional species of the *Passiflora vitifolia* complex: Amazonian species diversity due to edaphically differentiated communities. *Plant Systematics & Evolution*, **137**, 95–105.

Gentry, A.H. (1982) Neotropical floristic diversity: phytogeographical connections between Central and South America, Pleistocene climatic fluctuations, or an accident of the Andean orogeny? *Annals of the Missouri Botanical Garden*, **69** (3), 557–593.

Gentry, A.H. (1988) Changes in plant community diversity and floristic composition on environmental and geographic gradients. *Annals of the Missouri Botanical Garden*, **75** (1), 1–34.

Hamilton, S.K., Kellndorfer, J., Lehner, B. and Tobler, M. (2005) Remote sensing of floodplain geomorphology as a surrogate for biodiversity in a tropical river system (Madre de Dios , Peru). *Geomorphology*, **89**, 23–38.

Harrison, S. and Grace, J.B. (2007) Biogeographic affinity helps explain productivity-richness relationships at regional and local scales. *The American naturalist*, **170**(Suppl), S5–15.

Hoorn, C., Wesselingh, F.P., ter Steege, H. *et al.* (2010) Amazonia through time: andean uplift, climate change, landscape evolution, and biodiversity. *Science*, **330** (6006), 927–931.

Householder, J.E., Janovec, J.P., Tobler, M.W. *et al.* (2012) Peatlands of the Madre de Dios River of Peru: distribution, geomorphology, and habitat diversity. *Wetlands*, **32** (2), 359–368.

Householder, J.E., Wittmann, F., Tobler, M. and Janovec, J.P. (2015) Montane bias in lowland Amazonian peatlands: plant assembly on heterogenous landscapes and potential significance to palynological inference. *Palaeogeography, Palaeoclimatology, Palaeoecology*, **423**, 138–148.

Jackson, M. and Colmer, T. (2005) Response and adaptation by plants to flooding stress. *Annals of Botany*, **96**, 501–505.

Junk, W.J., Fenandez Piedade, M.T., Schöngart, J. *et al.* (2011) A classification of major naturally occurring Amazonian lowland wetlands. *Wetlands*, **31**, 623–640.

Kalliola, R., Puhakka, M., Salo, J. *et al.* (1991) The dynamics, distribution and classification of swamp vegetation in Peruvian Amazonia. *Annales Botanici Fennici*, **28** (3), 225–239.

Kubitzki, K. (1989) The ecogeographical differentiation of Amazonian inundation forests. *Plant Systematics and Evolution*, **162**, 285–304.

Laliberté, E. and Legendre, P. (2010) A distance-based framework for measuring functional diversity from multiple traits. *Ecology*, **91** (1), 299–305.

Marchant, R., Almeida, L., Behling, H. *et al.* (2002) Distribution and ecology of parent taxa of pollen lodged within the Latin American Pollen Database. *Review of Palaeobotany and Palynology*, **121**, 1–75.
Oliveira-Filho, A. and Ratter, J. (1995) A study of the origin of central Brazilian forests by the analysis of plant species distribution patterns. *Edinburgh Journal of Botany*, **52**, 141–194.
Parolin, P., Junk, W. and Piedade, M.T.F. (2000) Seed mass in Amazonian floodplain forests with contrasting nutrient supplies. *Journal of Tropical Ecology*, **16** (3), 417–428.
Parolin, P., de Simone, O., Haase, K. *et al.* (2004) Central Amazonian floodplain forests: tree adaptations in a pulsing system. *The Botanical Review*, **70** (3), 357–380.
Phillips, O. and Miller, J. (2002) *Global Patterns of Plant Diversity: Alwyn H. Gentry's Forest Transect Data*, Missouri Botanical Garden Press.
Pillon, Y., Munzinger, J., Amir, H. and Lebrun, M. (2010) Ultramafic soils and species sorting in the flora of New Caledonia. *Journal of Ecology*, **98** (5), 1108–1116.
Pitman, N.C.A., Guevara Andino, J.E., Aulestia, M. *et al.* (2014) Distribution and abundance of tree species in swamp forests of Amazonian Ecuador. *Ecography*, **37** (9), 902–915.
Prance, G.T. (1979) Notes on the vegetation of Amazonia III. The terminology of Amazonian forest types subject to inundation. *Brittonia*, **31** (1), 36–38.
Punyasena, S.W., Eshel, G. and McElwain, J.C. (2008) The influence of climate on the spatial patterning of neotropical plant families. *Journal of Biogeography*, **35**, 117–30.
Reich, P.B., Walters, M.B. and Ellsworth, D.S. (1997) From tropics to tundra: global convergence in plant functioning. *Proceedings of the National Academy of Sciences of the United States of America*, **94** (25), 13730–13734.
Ruokolainen, K., Schulman, L. and Tuomisto, H. (2001) On Amazonian peatlands. *International Mire Conservation Group Newsletter*, **4**, 8–10.
Sepkoski, J. and Kendrick, D. (1993) Numerical experiments with model monophyletic and paraphyletic taxa. *PaleoBiology*, **19**, 168–184.
Steyermark, J. (1979) Flora of the Guayana Highland: endemicity of the generic flora of the summits of the Venezuela Tepuis. *Taxon*, **28**, 45–54.
Tanner, E., Vitousek, P.M. and Cuevas, E. (1998) Experimental investigation of nutrient limitations of forest growth on wet tropical mountains. *Ecology*, **79** (1), 10–22.
Tuomisto, H. and Dalberg Poulsen, A. (1998) Edaphic distribution of some species of the fern genus Adiantum in Western Amazonia. *Biotropica*, **30**, 392–99.
Wittmann, F., Schöngart, J., Montero, J.C. *et al.* (2006) Tree species composition and diversity gradients in white-water forests across the Amazon basin. *Journal of Biogeography*, **33** (8), 1334–1347.
Wittmann, F., Householder, J.E., Piedade, M.T.F. *et al.* (2013) Habitat specificity, endemism and the neotropical distribution of Amazonian white-water floodplain trees. *Ecography*, **36** (6), 690–707.

8

Forest Composition and Spatial Patterns across a Western Amazonian River Basin: The Influence of Plant–Animal Interactions

Varun Swamy

Abstract

Although much effort has been devoted to studying the composition and spatial structure of tropical tree communities and the underlying processes, the status quo remains a lack of consensus. Common issues in most previous studies include a lack of true replication, and an exclusive focus either on a single cohort or purely *intra*-cohort spatial patterns. Here, I use data from a set of forest dynamics plots spread across a *ca.* 80,000 sq km western Amazonian river basin that includes stems of three distinct size/age cohorts to examine compositional and spatial patterns of the woody plant community. Salient compositional patterns include an 'oligarchy' comprised of a small number of high-abundance taxa with a long tail of low-abundance taxa, and an overall predominance of dioecious and endozoochorous taxa. Analyses of spatial patterns indicate an overall pattern of 'displaced self-replacement', which confirms the outcome of ecological interactions predicted by the Janzen-Connell hypothesis.

Keywords *tropical forests; lowland Amazon rainforests; Western Amazonia; Madre de Dios Basin; Peru; plant-animal interactions; seed dispersal; forest regeneration; tree diversity; Janzen-Connell hypothesis; spatial organization; displaced self-replacement*

8.1 Introduction

Since the latter half of the 20th century, an enormous amount of research has been conducted on the factors that influence the dynamics of plant regeneration in lowland tropical rainforests, resulting in several major paradigms that relate regeneration patterns and the underlying processes. Given the multitudes of factors and the time involved in the journey from fruit to reproductive adult, it is unsurprising that the scientific literature is comprised of vastly differing paradigms of processes and phenomena considered most significant in their influence on the maintenance of species diversity and the spatial organization of tropical forest tree communities. Perhaps the only point of agreement amongst these paradigms is that forest regeneration takes a long time and involves

Forest structure, function and dynamics in Western Amazonia, First Edition.
Edited by Randall W. Myster.

multiple ontogenetic phases, each of which can be influenced by a plethora of biotic and abiotic factors. Prominent theories include niche pre-emption (Whittaker 1965), "escape in space" from natural enemies (Connell 1971; Janzen 1970), differentiation in the regeneration niche (Grubb 1977), gap dynamics (Denslow 1987; Hartshorn 1978), intermediate disturbance (Connell 1978), community drift (Hubbell 1979), lottery competition (Chesson and Warner 1981), negative density dependence (Condit *et al.* 1994; Harms *et al.* 2000; Hubbell *et al.* 1990; Hubbell and Foster 1986; Wills *et al.* 1997), spatial heterogeneity of limiting resources (Pacala and Tilman 1994), dispersal limitation (Hurtt and Pacala 1995; Tilman 1994) and recruitment limitation (Hubbell *et al.* 1999; Muller-Landau *et al.* 2002; Wright 2002). These theories span the gamut in their emphases on the relative importance of biotic vs. abiotic factors, mechanisms characterized by stochasticity vs. determinism, and in spatio-temporal resolution from sub-hectare to global, and days to millennia.

Only one of these theories – commonly referred to as the "Janzen-Connell (J-C) hypothesis" (Connell 1971; Janzen 1970) – describes a primary role for animals in plant regeneration processes. This is surprising given that animals play crucial roles at virtually every step in plant reproduction and recruitment in tropical rainforests: pollination, seed dispersal, seed and seedling predation, and seedling growth and maturation. Consequently, any human-induced alteration of the faunal community within an ecosystem can be expected to disrupt the processes that control the regeneration of its plant community. Tropical forests are particularly vulnerable to such disruptions because of their high floral and faunal diversity, and the extent and complexity of plant–animal interactions in these ecosystems. In lowland tropical forests of Western Amazonia, large primates such as black spider monkeys (*Ateles chamek*) and wooly monkeys (*Lagothrix lagothricha*) have been identified as critically important seed dispersers, owing to their high densities, large ranges and almost entirely frugivorous diets that include the fleshy fruits of a significant proportion of canopy tree species (Andresen 1999; Terborgh 1983). Other important endozoochorous vertebrate dispersers of large-seeded, fleshy-fruited canopy trees in Western Amazonian rainforests include capuchin monkeys (*Cebus* and *Sapajus spp.*), large birds such as guans (*Penelope, Aburria* and *Pipile* spp.) and trumpeters (*Psophia* sp.), and terrestrial vertebrates such as the South American tapir (*Tapirus terrestris*). By ingesting mature fruits in the canopy or at the base of fruiting trees and later defecating or regurgitating the intact pulp-free seeds away from their parent trees, vertebrate frugivores are in effect ensuring the "escape in space" of these seeds through the avoidance of consumption or infestation by a variety of host-specific seed/seedling predators and pathogens that operate in areas of high seed/seedling density around the crowns of reproductive adults, as described by the J-C hypothesis. These dispersed seeds are far more likely to survive, germinate and produce sapling recruits that represent the future generation of adult trees in the rainforest. Frugivorous vertebrates are thus critical for forest regeneration and maintaining plant diversity and spatial organization in lowland tropical rainforests.

The spatial and compositional patterns of tropical forest tree communities and the underlying processes that produce them have attracted scientific inquiry since the late 19th century (Wallace 1878) and have been subjected to extensive theoretical and empirical research over the past half century (Condit *et al.* 2000; Hubbell 1979; Hubbell *et al.* 1999; Peters 2003; Wills *et al.* 1997). Beginning in the 1980s, the establishment of large forest dynamics plots in several tropical forest sites across the world provided sample sizes adequate to examine this question at sufficiently large spatial scales. Over this period, the prevailing notion of tropical forest spatial organization has veered across the spectrum from "hyper-spaced" (Richards 1952; Wallace 1878) to "clumped" (Condit *et al.* 2000; Hubbell 1979), with multiple intermediate, conditional and increasingly nuanced scenarios based on life history traits (Mangan *et al.* 2010; Plotkin *et al.* 2000; Seidler and Plotkin 2006). However, the status quo remains a lack of consensus. A common limitation/oversight of most previous studies of spatial patterns in tropical tree communities has been an exclusive focus either on a single size/age cohort, or a comparison of intra-cohort spatial organization amongst different size/age classes; in both cases, spatial relationships between cohorts have been largely ignored. This is a significant issue because inter-cohort interactions are the basis of a widely supported mechanism for the maintenance of diversity and spatial structure in tropical tree communities – the J-C hypothesis. At the level of the focal tree or species, the J-C hypothesis predicts a recruitment pattern skewed away from conspecific adult trees because of near-total recruitment failure under/around the crowns of conspecific adults, caused primarily by density/distance-responsive host-specific natural enemies. Integrated across the entire tree community, this would result in a landscape-scale pattern of "displaced self-replacement", such that the offspring cohort as a whole is displaced away from its adult tree cohort as a whole. This notion of displacement across generations is seemingly in direct contrast with the observation of spatial aggregation or clumped distributions that has been noted in previous studies, but it is not necessarily contradictory. Intra-cohort clumping does not preclude inter-cohort displacement; to a certain extent, the J-C hypothesis predicts both outcomes. Intra-cohort aggregation can be produced by the accumulation of dispersed seeds in "safe sites" that are limited in availability in the forest stand and likely skewed in their spatial distribution based on the spatial configuration of conspecific adult trees. In addition, environmental filtering due to species-specific abiotic regeneration niches can further skew the spatial distribution of young stem recruits, resulting in marked intra-cohort aggregation for any given species. Another unresolved issue is how this intra-cohort aggregation changes over time, from younger to older stem recruits, and eventually to reproductive-sized adult trees. One scenario is a decrease in intra-cohort aggregation with ontogenetic progression from young recruits to adult trees – which would suggest a primary role for density-dependent resource competition; whereas the opposite scenario – for example an increase in intra-cohort aggregation from young recruits to reproductive-sized adult trees – would support a prominent role for environmental filtering, with recruits surviving largely in optimal niche conditions to become adult trees.

A further limitation of most previous studies of spatial patterns in tropical tree communities has been a lack of true replication – several studies either have used a single albeit large sample, or have analyzed tree plot data from multiple study sites with little or no overlap in species. This greatly limits the scope of inference of results, and highlights the need for multiple independent samples drawn from a sufficiently large area and from a common regional species pool in order to extend the scope of inference beyond a single site.

Here, I use data from a set of forest dynamics plots spread across a *ca.* 80,000 km^2 Western Amazonian river basin that includes stems of three distinct size/age cohorts to examine compositional and spatial patterns of the woody plant community. I hypothesize that the spatial organization of conspecific stems will demonstrate significant *intra*-cohort aggregation, i.e., "clumping" (hypothesis 1), which will decrease in intensity with increasing size/age as a result of resource-based density-dependent mortality (hypothesis 2). *Inter*-cohort spatial organization of conspecific stems, on the other hand, will demonstrate significant "dissociation" between generations (hypothesis 3). Overall, intra- and inter-cohort spatial organization will combine to produce a pattern of "displaced self-replacement" (hypothesis 4).

8.2 Methods

8.2.1 Site description and history

This study uses data from four long-term forest dynamics plots located within the *ca.* 85,000 km^2 Madre de Dios River Basin (MDDB) in southeastern Peru (Figure 8.1, Table 8.1). The basin receives 2500–3500 mm of rainfall annually,

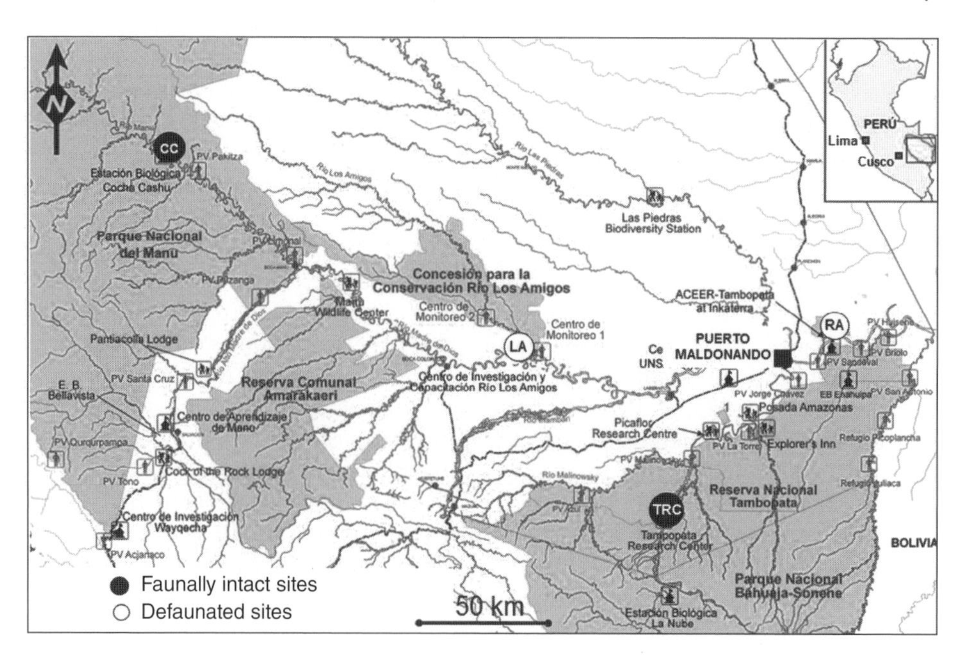

Figure 8.1 Map of MDDB, Peru, showing locations of four long-term forest dynamics plots used in this study and their faunal status. Original base map created by Nelson Gutiérrez, Amazon Conservation Association (2007).

Table 8.1 Summary data of four forest dynamics plots used in this study.

Site (site code)	Location	Data collection		Disperser density[1] (Individuals km^{-2})		White lipped peccary abundance
		Tree plot	Recruitment plot	Large primates	Large birds	
Cocha Cashu (CC)	11° 52′ S, 71° 24′ W	2009	2009	37.5	29.2	High
Tambopata Research Center (TRC)	13° 7′ S, 69° 36′ W	2008	2009	15.9	44.0	High
Los Amigos (LA)	12° 34′ S, 70° 4′ W	2008	2009	3.9	18.2	Intermediate
Reserva Amazonica (RA)	12° 32′ S, 69° 3′ W	2008	2009	0.0	8	~Absent

with a pronounced dry season between June and September (Gentry and Terborgh 1990; Tobler *et al.* 2009). The MDDB contains vast expanses of legally protected natural habitat that have been minimally impacted by recent human activity, including Manu National Park, Tambopata National Reserve and Bahuaja-Sonene National Park, which are amongst the world's most species-rich biodiversity hotspots. The MDDB is also home to tens of thousands of inhabitants, comprised of several indigenous groups as well as Andean immigrants engaged in natural resource-exploitative activities such as logging and mining. The main biodiversity threats in this region include deforestation through logging, mining, road-building and agricultural activities, as well as intensive hunting of large-bodied vertebrates (Ohl-Schacherer *et al.* 2007; Shepard *et al.* 2010; Terborgh 1999). Recent studies in the region indicate that large-bodied frugivores including primates (*A. chamek*, *L. lagothrica*, *Sapajus macrocephalus* and *Cebus albifrons*), birds (*Penelope*, *Pipile* and *Psophia* spp.) and terrestrial vertebrates (*Tayassu pecari*) are being exploited unsustainably, and populations of these animals are locally extinct or significantly depressed at intensively hunted sites (Nunez-Iturri and Howe 2007; Ohl-Schacherer *et al.* 2007; Terborgh *et al.* 2008).

The four sites (Figure 8.1, Table 8.1) are spread across the MDDB in a northwest-southeast orientation, including three of its major tributaries:

- Manu River: CC site, in the vicinity of Cocha Cashu Biological Station in the core of Manu National Park.
- Amigos River: LA site, in the vicinity of Los Amigos Biological Station, on the border of the Los Amigos Conservation Concession (MDDB) by the confluence of the Amigos and Madre de Dios rivers.
- Madre de Dios River: RA site, located within Reserva Amazonica, a private ecotourism concession situated two hours downstream from the city of Puerto Maldonado.

- Tambopata River: TRC site, in the vicinity of Tambopata Research Center, located at the boundary between Tambopata National Reserve and Bahuaja-Sonene National Park.

The minimum Euclidean distance between plots ranges from approximately 90 km (RA and TRC) to approximately 270 km (CC and RA). All sites are situated in mature floodplain forest that has never been subjected to commercial timber extraction. CC site is located within the *ca.* 20,000 km^2 Manu National Park where human-induced disturbances such as logging, agriculture and hunting have been absent since the establishment of the park in 1973. TRC site is located at the intersection of Tambopata National Reserve and Bahuaja-Sonene National Park, which together constitute *ca.* 17,000 km^2. No hunting has taken place within 40 km of TRC since the early 1990s, when the transitionary Tambopata-Candamo Reserve Zone was created and eventually led to the official designation of the two adjoining protected areas (Naughton-Treves *et al.* 2003). LA site was intensively hunted over a 15-year period (1982–1999) when the station served as the headquarters for a series of gold-mining operations (Pitman 2010). Unmonitored hunting of large vertebrates in the area continued until 2001, when a conservation concession was created by the Peruvian government to protect the watershed of the Amigos river. Since 2003, the concession has been monitored and protected by the Asociación para la Conservación de la Cuenca Amazonica (ACCA), a Peruvian non-governmental organization, and hunting is virtually non-existent (Tobler 2008; Tobler *et al.* 2009). The RA site was established as a private reserve in 1977; however, its reserve status was not renewed in the late 1980s and subsistence farmers have since colonized and converted significant segments of the riverine forest to small-scale agriculture. A small core area (*ca.* 1,500 hectares) remains a private reserve administered by an eco-tourism operation and contains structurally undisturbed mature forest. Hunting pressure has increased in the surrounding area in recent years (Kirkby *et al.* 2000).

8.2.2 Study design

Each site has a 200 m × 200 m (4 hectare) forest dynamics plot in which all stems ≥10 cm diameter at breast height (dbh) have been identified to species (or unique morpho-species with closest taxonomic designation possible) and tagged, measured (dbh to 0.1 cm) and mapped (longitude and latitude to 0.1 m) (Figure 8.2). In addition, the central hectare (100 m × 100 m) of each plot comprises a recruitment monitoring area in which all woody stems ≥1 m tall have been identified, tagged, measured and mapped. Recruit stems are divided into two cohorts based on size: "saplings" (≥1 m tall and <1 cm diameter) and "juveniles" (≥1 cm diameter and <10 cm dbh). These size thresholds correspond well with two distinct age cohorts: stems of less than 1 cm diameter (saplings) typically represent recruits less than 20 years old (with the exception of a small number of fast-growing species), whereas juveniles typically represent recruits ≥20 years old (Hubbell 2004; Welden *et al.* 1991). A third size/age cohort of "adult" trees comprised stems ≥10 cm dbh. However, 10 cm dbh represents an arbitrary threshold for reproductive adult status, and is clearly inappropriate for

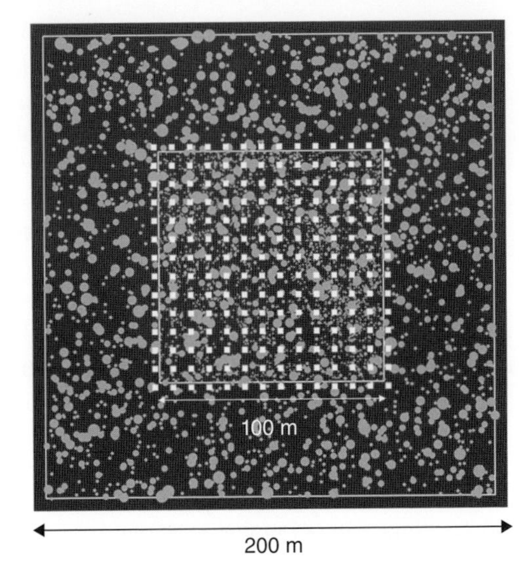

Figure 8.2 Layout and design of each of four study sites, showing 4-ha (200 m × 200 m) forest dynamics plot and centrally located seed fall monitoring grid (104 m × 104 m) and 1-ha (100 m × 100 m) recruitment monitoring area.

canopy-statured tree species. Therefore, the adult cohort used in subsequent spatial analyses was restricted to individuals likely to be of reproductive size, using a simple criterion: for species with a median dbh ≥30cm (canopy-statured), stems that were under the species-specific median dbh value were excluded.

8.3 Analysis

8.3.1 Compositional patterns

Compositional patterns were analyzed after classifying the stem inventory dataset into two distinct size/age-based vertical strata: the canopy and sub-canopy "tree" layer (all stems ≥10 cm dbh), and the understory "recruit" layer (all stems ≥1 m tall and <10 cm dbh). For each stratum, composition was analyzed at the family and species level for each of the four sites, and pooled across all four sites for a basin-scale inference.

8.3.2 Spatial patterns

8.3.2.1 Intra-cohort spatial patterns

Intra-cohort spatial organization was examined by calculating nearest-neighbor distances (NNDs) for conspecific stems within a cohort, and computing species-specific median NNDs separately for each cohort (saplings, juveniles, adults) at each site. Median NNDs were computed only for species with ≥3 stems per focal cohort per site.

To test whether the spatial distribution of conspecific stems within a cohort is distinct from that of heterospecific stems of the same recruitment cohort, a significant *intra*-cohort spatial signal, median NNDs of each focal species were compared with median NNDs of randomly sampled heterospecific stems of the same cohort in the same forest stand, as follows:

For each focal species within a cohort at each site:

- 1 iteration: computation of median NND for a randomly selected set of stems in the same cohort based on distance to nearest neighbor for each stem, with N (randomly selected stems) = N (stems of focal species).
- 1,000 such iterations were run, with median NND calculated in each case, and a percentile rank (p-rank) of the observed median NND of each focal species was computed based on the 1,000 simulated median NNDs from random selections of stems.

The spatial pattern of the observed median NND of each focal species (per-cohort and per-site) was classified as follows:

- "Clumped" or "Aggregated": p-rank <0.025 (within lowest 2.5%) "Dissociated" or "Hyper-spaced": p-rank >0.975 (within highest 2.5%) "Random" or "Neutral": 0.025 < p-rank <0.975 (within 95% of simulated median NND values)

8.3.2.2 Inter-cohort spatial patterns

Inter-cohort spatial organization was examined by calculating "recruitment distance" (RD), defined as the distance between a focal sapling and its nearest conspecific adult tree, and computing the median RD on a per-species per-site basis. Median RDs were computed only for species with ≥1 adult tree and ≥3 saplings per site.

To test whether the spatial distribution of conspecific saplings and adult trees is distinct from that of heterospecific saplings and adult trees – a significant *inter*-cohort spatial signal, median RDs of each focal species were compared with median RDs of randomly sampled heterospecific saplings and adult trees within the same forest stand, as follows:

For each focal species at each site:

- 1 iteration: computation of median RD for a randomly selected set of saplings and adult trees within the same forest stand, based on RD for each focal sapling, with N (randomly selected sapling and adult stems) = N (sapling and adult stems of focal species).
- 1,000 such iterations were run, with median RD calculated in each case, and a percentile rank (p-rank) of the observed median RD of each focal species was computed based on the 1,000 simulated median RDs from random selections of sapling and adult stems. The spatial pattern of the observed median RD of each focal species at each site was classified as follows:
- "Clumped" or "Aggregated": p-rank <0.025 (within lowest 2.5%) "Dissociated" or "Hyper-spaced": p-rank >0.975 (within highest 2.5%) "Random" or "Neutral": 0.025 <p-rank <0.975 (within 95% of simulated RD values)
- All spatial analyses were performed using the 'spatstat' package in R version 3.2.1.

8.4 Results

8.4.1 Compositional patterns

The dataset comprised 9,253 tree-sized individuals (≥10 cm dbh) sampled across a total of 16 hectares, and 38,099 understory recruit stems sampled across a total

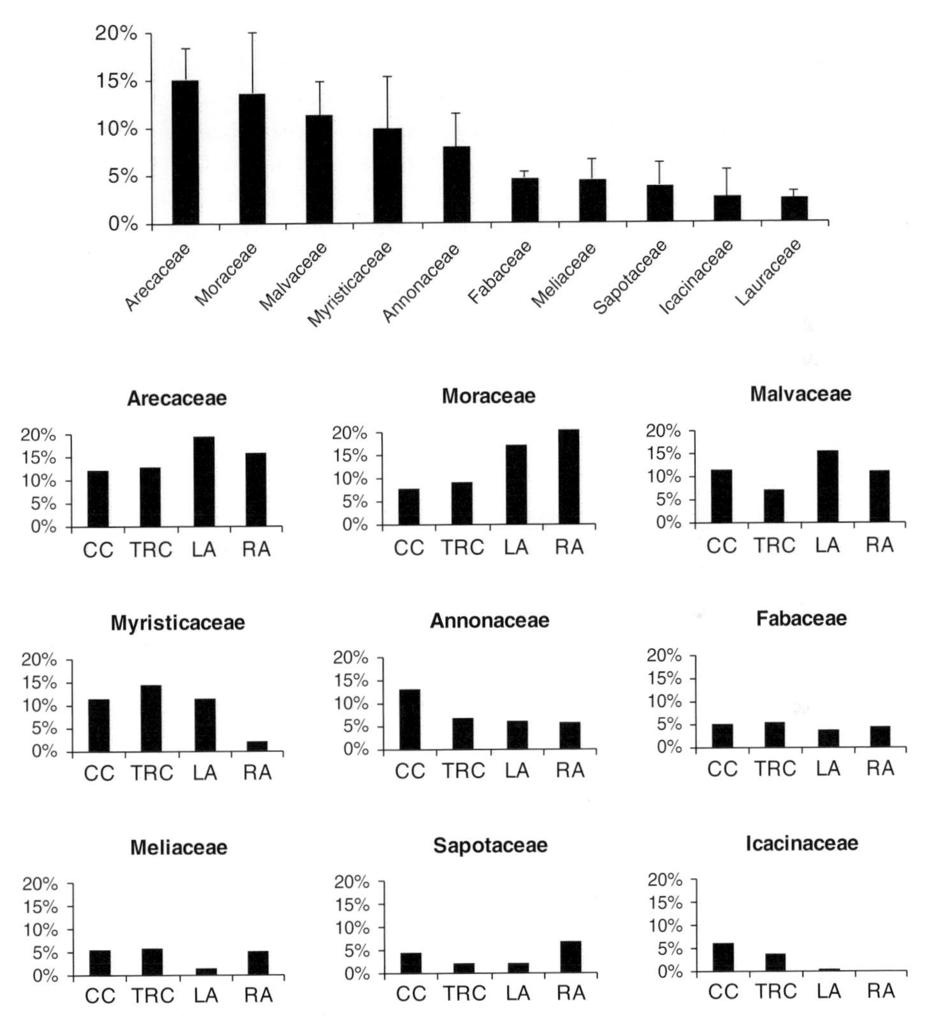

Figure 8.3 Mean percentage of all stems comprised by the 10 most abundant families of trees (stems ≥10 cm diameter) in four 4-ha forest dynamics plots spread across the MDDB, Peru: pooled across sites (±SD), and within each site for first 9 families.

of 4 hectares. Trees were comprised of 57 families and more than 600 species, and understory recruits were comprised of 80 families and more than 1,000 species.

The ten most abundant families of trees across the MDDB are Arecaceae, Moraceae, Malvaceae, Myristicaceae, Annonaceae, Fabaceae, Meliaceae, Sapotaceae, Icacinaceae and Lauraceae (Figure 8.3). Arecaceae – the palm family – is notably and consistently dominant across sites, comprising up to 19% and no less than 12% of stems within a site, with a basin-scale mean of 15%. None of the next nine most abundant families topped more than 15% at any given site or averaged across the basin, except for Moraceae at one (RA) site. Some families exhibit considerable inter-site variation in abundance, for example Moraceae, Malvaceae and Myristicaceae, whereas others are notably consistent in their abundance across sites, for example Fabaceae and Lauraceae.

At the species level, the 10 most abundant trees across the MDDB were *Iriartea deltoidea* (Arecaceae), *Otoba parvifolia* (Myristicaceae), *Pseudolmedia laevis* (Moreaceae), *Quararibea wittii* (Malvaceae), *Calatola venezuelana* (Icacinaceae), *Theobroma cacao* (Malvaceae), *Oxandra acuminata* (Annonaceae), *Astrocaryum murumuru* (Arecaceae), *Guarea macrophylla* (Meliaceae) and *Celtis schippii* (Cannabaceae) (Figure 8.4). *I. deltoidea* is dominant at 3 of 4 sites

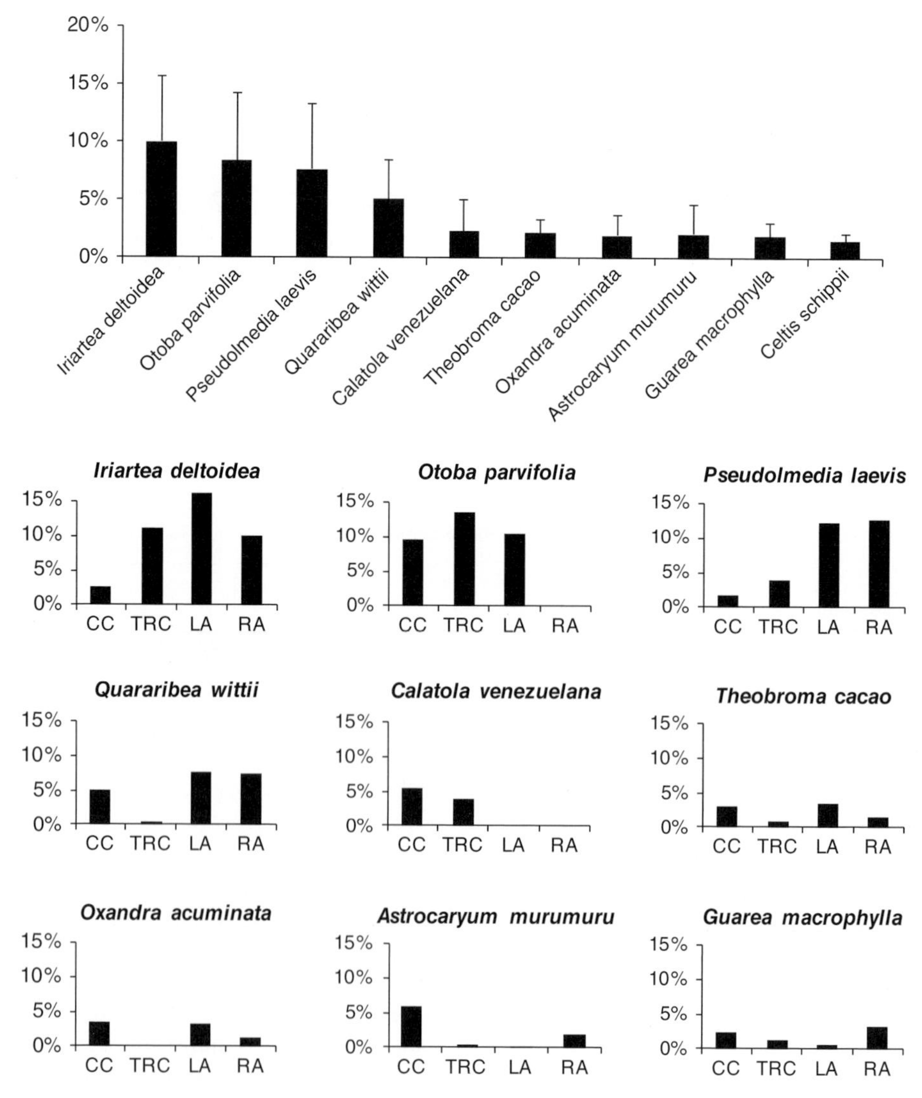

Figure 8.4 Mean percentage of all stems comprised by the 10 most abundant species of trees (stems ≥10 cm diameter) in four 4-ha forest dynamics plots spread across the MDDB, Peru: pooled across sites (±SD), and within each site for first 9 species

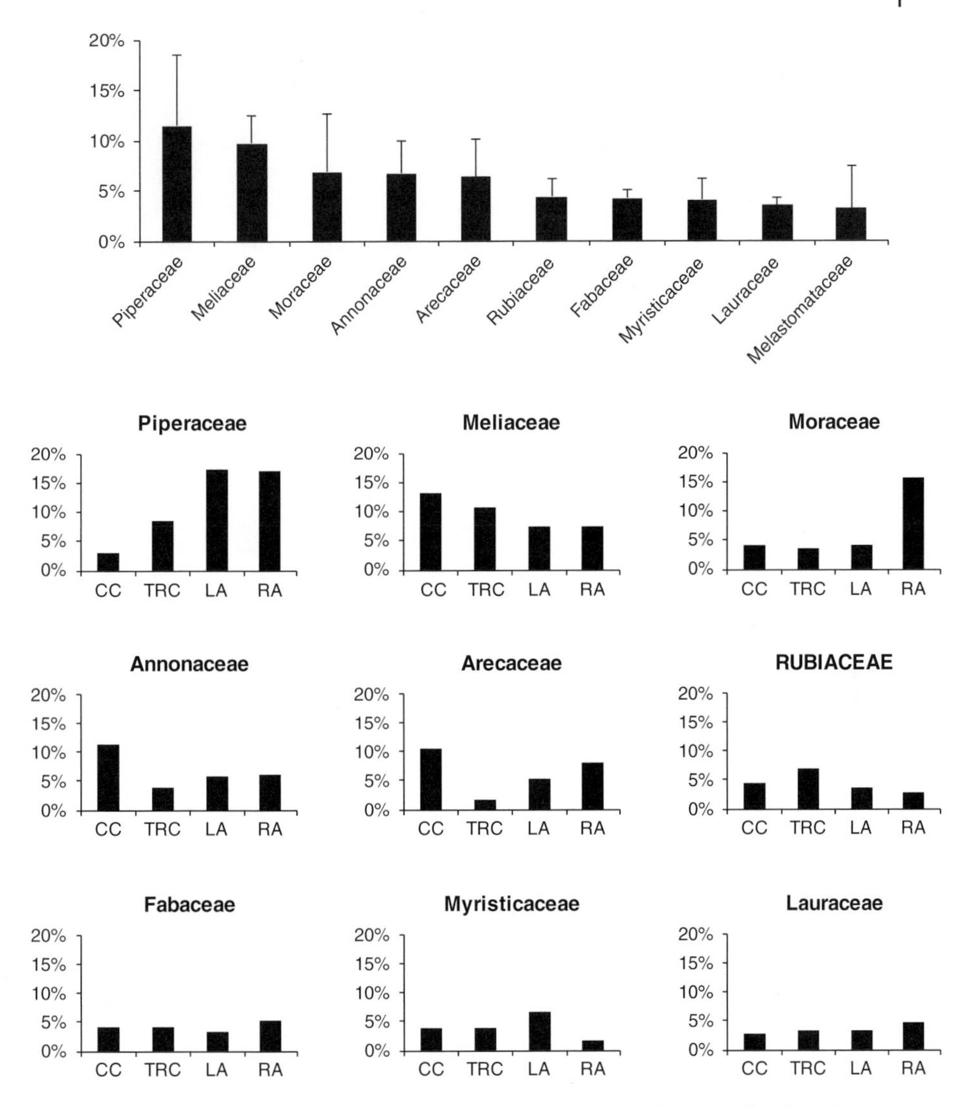

Figure 8.5 Mean percentage of all stems comprised by the 10 most abundant families of understory woody taxa (stems ≥1 m height, <10 cm diameter) in the central hectare of 4 forest dynamics plots spread across the MDDB, Peru: pooled across sites (±SD), and within each site for first 9 families.

(10–16% of all stems), as is *O. parvifolia* (10–14%), whereas the next 8 most abundant species vary considerably in abundance between sites.

Amongst understory woody taxa, the 10 most abundant families of trees across the MDDB were Piperaceae, Meliaceae, Moraceae, Annonaceae, Arecaceae, Rubiaceae, Fabaceae, Myristicaceae, Lauraceae and Melastomataceae (Figure 8.5). Piperaceae comprised up to 17% of all stems within a site (LA and RA), and Moraceae dominated one site (RA), comprising 16% of all stems. Fabaceae and Lauraceae are notably consistent in their abundance across sites, whereas all the other eight most dominant families exhibit considerable inter-site variation in abundance.

At the species level, the 10 most abundant understory stems across the MDDB are *Piper laevigatum* (Piperaceae), *Guarea kunthiana* (Meliaceae), *P. laevis* (Moraceae), *Piper glabratum* (Piperaceae), *O. parvifolia* (Myristicaceae), *G. macrophylla* (Meliaceae), *Miconia triplinervis* (Melastomataceae), *Chrysochlamys ulei* (Clusiaceae), *Piper hispidum* (Piperaceae) and *A. murumuru* (Arecaceae) (Figure 8.6). *P. laevigatum* and both *Guarea* species are relatively

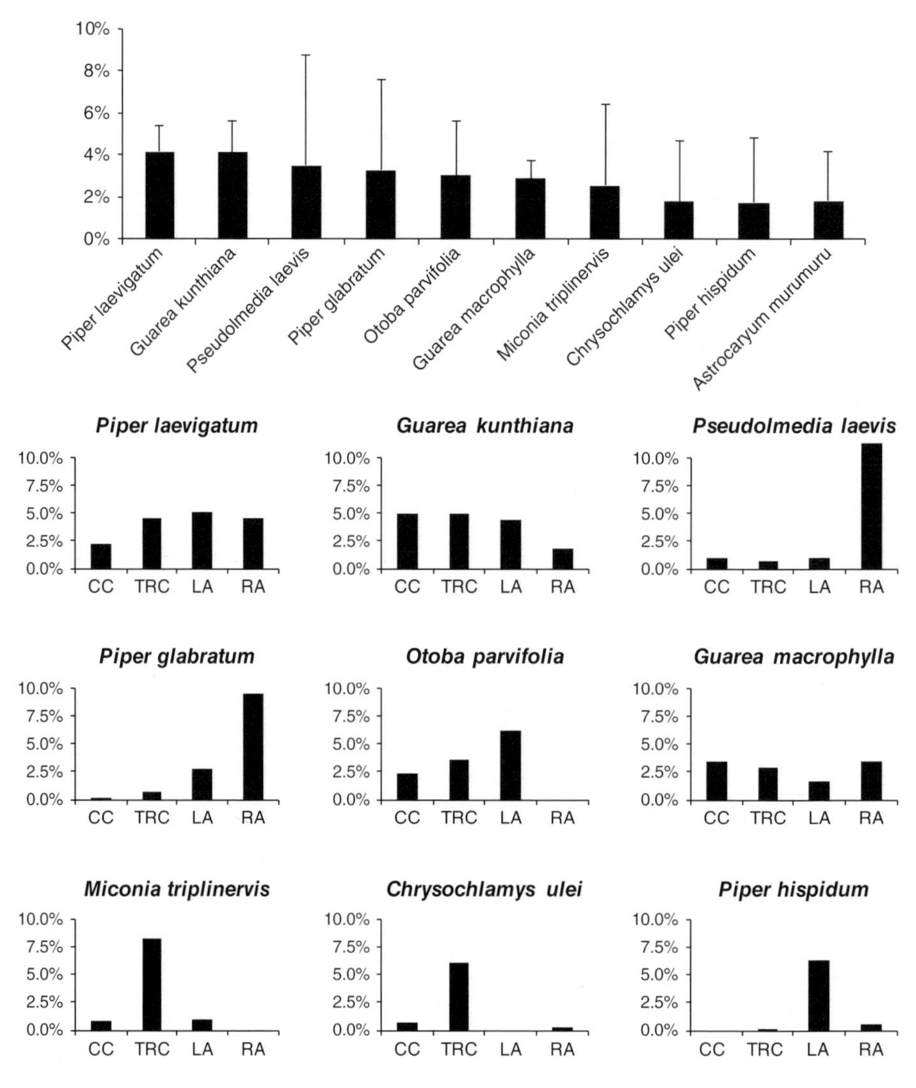

Figure 8.6 Mean percentage of all stems comprised by the 10 most abundant species of understory woody taxa (stems ≥1 m height, <10 cm diameter) in the central hectare of 4 forest dynamics plots spread across the MDDB, Peru: pooled across sites (±SD), and within each site for first 9 species.

consistent in their abundance across sites, whereas multiple species were strikingly patchy in their distribution, with very high abundance at a single site (*P. laevis* at RA, *C. ulei* at TRC and *P. hispidum* at LA) and low abundance at another three sites.

8.4.2 Spatial patterns

8.4.2.1 Intra-cohort spatial patterns

NNDs were computed and analyzed for a cumulative 1,109 taxa pooled across cohorts and sites, representing 239 unique species (194 adult trees, 219 juvenile and 176 sapling taxa). Twelve species were represented in all 3 cohorts at all 4 sites, and 36 species were represented in at least 9 out of a maximum of 12 possible site × cohort combinations.

Median NND values averaged across sites ranged from 14.4–16.6 m for juveniles, 17.1–21 m for saplings, and 34.2–37 m for adult trees (Figure 8.7). Median NND calculations for saplings and juveniles are based on the same spatial scale (100 m × 100 m), whereas the considerably larger adult NNDs are calculated across a 200 m × 200 m spatial scale. Therefore, only sapling and juvenile NNDs are directly comparable. Taking this into account, sapling NNDs were consistently larger than juvenile NNDs at all four sites (2–6.6 m) and pooled across sites (4 m) at the basin scale.

Percentile ranks of observed median NNDs, in comparison with computations based on 1,000 iterations of randomly sampled stems of the same cohort, revealed that 16% of species across all three cohorts combined have a significantly clumped spatial distribution, whereas 82% of species confirmed to a random or neutral spatial distribution, that is conspecific stems are no more or less aggregated than randomly sampled heterospecific stems (Figure 8.8). Only 2% of all species, pooled across cohorts, showed a dissociated intra-cohort spatial distribution. These proportions based on pooled results from all three

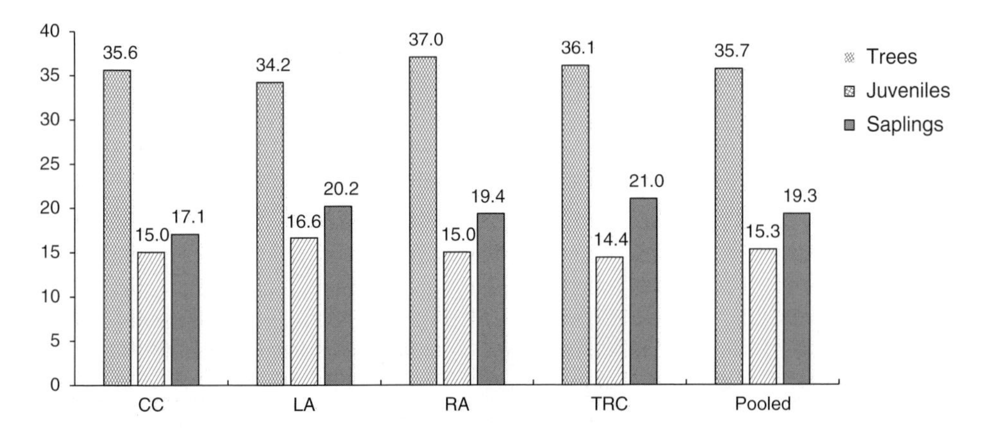

Figure 8.7 Median NND (nearest neighbor distance, m), i.e. distance between a stem and its nearest conspecific neighbor within a cohort, averaged across species for three distinct size-age stem cohorts at each of four long-term forest dynamics plots spread across the MDDB, Peru, and pooled across sites.

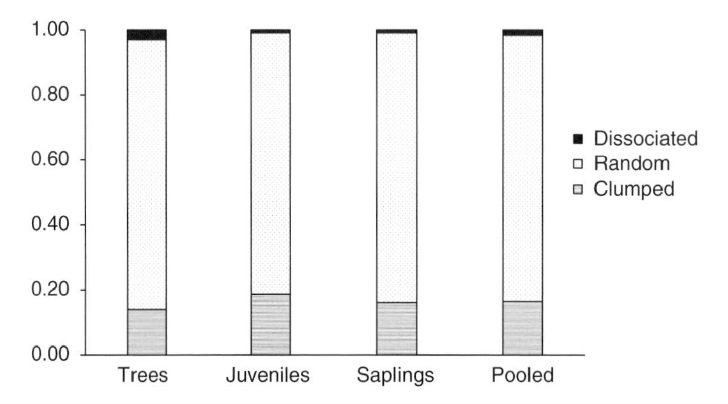

Figure 8.8 Proportions of species whose observed median NND (nearest neighbor distance of conspecific stems) confirmed to a "clumped", "random", or "dissociated" intra-cohort spatial pattern based on its percentile rank compared to 1,000 iterations of randomly sampled heterospecific stems of the same cohort. Results shown separately for each cohort and pooled across cohorts, in each case pooled across all four sites.

cohorts were also reflected within each cohort (14–16% clumped, 80–83% neutral, 1–3% dissociated). Median percentile ranks of observed NNDs were comparable between the three stem cohorts, reflecting a modest increase with ontogenetic progression from sapling to adult tree (saplings: 0.227 ± 0.016, juveniles: 0.253 ± 0.014, adult trees: 0.276 ± 0.016).

8.4.3 Inter-cohort spatial patterns

RDs were computed and analyzed for a cumulative 257 taxa pooled across cohorts and sites, representing 160 unique species (CC: 76, TRC: 66, LA: 58, RA: 57). Fourteen species were represented at all 4 sites, 29 species at 3 out of 4 sites, and 54 species in at least 2 of the 4 sites.

Median RD values for individual species ranged from 1.8–127.3 m. Averaged across species, per-site median RD values ranged from 39.1–50.9 m, with a basin-scale average of 43 m (median 35 m, mode 49 m) pooled across sites (Figure 8.9). This raw distance value of 43 m, in a more ecological context, equates to 2–10 multiples of the average crown radius of focal tree species; less than 2% of saplings are located within the crown zone of conspecific adult trees.

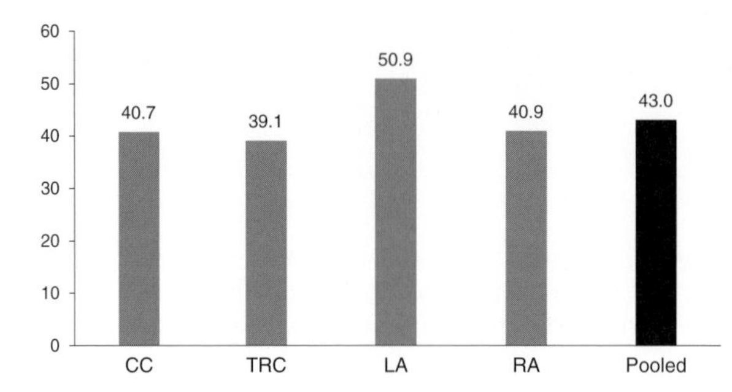

Figure 8.9 Median recruitment distance (RD, m), i.e. distance between a focal sapling and its nearest conspecific adult tree, averaged across species at each of four long-term forest dynamics plots spread across the MDDB, Peru, and pooled across sites.

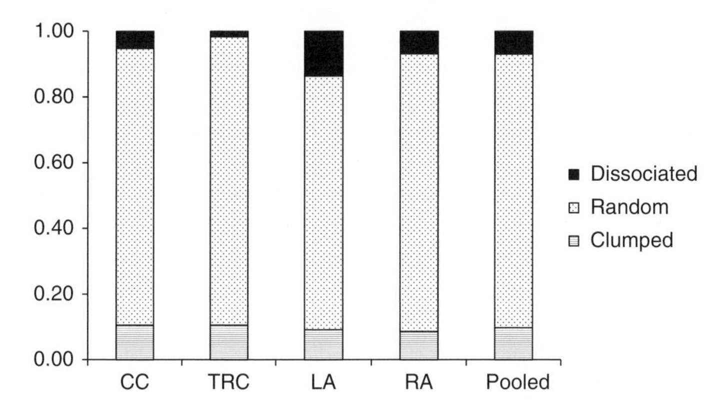

Figure 8.10 Proportions of species whose observed median recruitment distance confirmed to a "clumped", "random" or "dissociated" inter-cohort spatial pattern based on its percentile rank compared to 1,000 iterations of randomly sampled sapling and adult tree stems. Results shown separately for each site and pooled across all four sites.

Percentile ranks of observed median RDs, in comparison with computations based on 1,000 iterations of randomly sampled sapling and adult stems, revealed that 10% of all species, pooled across all four sites, have a significantly clumped inter-cohort spatial distribution, whereas 83% of species confirmed to a random or neutral spatial distribution, showing that saplings are no more or less aggregated around conspecific adult trees than random samplings of sapling and adult stems (Figure 8.10). Seven percent of all species showed a dissociated inter-cohort spatial distribution. These proportions based on pooled results across all four sites were also largely reflected within each site (9–11% clumped, 77–88% neutral, 2–14% dissociated); only the LA site had a notably larger proportion of species (14%) with a dissociated inter-cohort spatial distribution.

8.5 Discussion

Forest composition patterns revealed in these analyses support the notion of an "oligarchy" comprised of a small proportion of "hyperdominant" species and a long "tail" of species occurring at low densities, which has been previously reported in regional and continental scale studies (Pitman *et al.* 1999; ter Steege *et al.* 2013; Terborgh *et al.* 2002). However, these results are the first that are based on simultaneously examining compositional patterns of the tree and understory strata sampled in the same location, at multiple sites spanning a *ca.* 80,000 km^2 Western Amazonian river basin. This is a significant advancement over previous efforts, because the understory of Amazonian forests has been very poorly sampled to date, apart from the Center for Tropical Science (CTFS) network's 25-ha forest dynamics plots at Yasuní, Ecuador and Manaus, Brazil; even these plots only include stems ≥1 cm diameter, whereas the understory compositional data presented here include an additional younger recruit cohort of stems ≥1 m tall and <1 cm diameter.

Results from this study indicate that although canopy and understory strata share similar dominance-diversity patterns, they have relatively little overlap in terms of compositional similarity. Certain canopy and sub-canopy statured species are abundant in the understory, representing the "advance regeneration" guild of shade-tolerant long-lived tree species, for example *O. parvifolia, P. laevis, G. macrophylla,* etc. However, the understory woody flora is dominated by the shrub and "treelet" growth form exemplified by taxa such as *Piper, Miconia, Rinorea, Neea,* etc. and multiple Myrtaceae genera, which are never encountered in the ≥10 cm dbh size class that represents the (sub-)canopy stratum. The substantially higher species richness of the understory flora compared to the canopy layer is striking, even after accounting for difference in sample size of stems and area sampled. Most studies of tropical forest vegetation dynamics focus only on canopy-sized trees or use the rather arbitrary 10 cm dbh cut-off; these results suggest that ignoring the understory stratum significantly reduces the overall diversity of the woody plant community of Western Amazonian forests, and provides an incomplete, unrepresentative picture of their vegetation composition and diversity.

The importance of sufficiently large replicate samples distributed across the basin is illustrated by the strikingly patchy distribution of certain abundant species. For example, *O. parvifolia* is overall the second-most abundance species and is dominant at the CC, TRC and LA sites, but is completely absent at the RA site for no apparent reason since this site otherwise has a high species overlap with the CC site, which it is the farthest away from. Similarly, *Q. wittii* has strikingly low abundance at TRC compared to the other three sites, and *C. venezuelana* is entirely missing from the LA and RA sites. The patchiness of these species appears unrelated to geography since the sites in which they are most abundant are often the farthest apart. In the understory layer, patchy distribution of some of the abundant species might be more of a sampling artifact given the relatively small area (1 ha), or possibly the combined result of limited dispersal, microhabitat preferences and a recent recruitment spike due to favorable conditions in the understory such as light gaps, for example *C. ulei* and *M. triplinervis* at TRC and *P. hispidum* at LA.

Two other interesting results emerge from the analysis of compositional patterns: the predominance of dioecious species, and of endozoochory as the dominant dispersal syndrome of Western Amazonian woody flora. Of the 10 most abundant tree families, Moraceae (except the genus *Ficus*) and Myristicaceae taxa are ≈100% dioecious, as is a large fraction of Arecaceae taxa. The second- and third-most abundant species, *O. parvifolia* and *P. laevis,* are both dioecious. The ecological significance is that a significant and virtually unknown proportion of reproductive-sized trees in the forest stand will not produce a single fruit over the course of their lifetime, which can span multiple centuries; instead, their contribution to reproduction and regeneration is limited to gene flow via pollen transfer. This also has significant ecological implications for the carrying capacity of these forests in terms of the production of fleshy fruits that provide the nutritional needs of a diverse array of vertebrate and invertebrate frugivores and omnivores (Diaz-Martin *et al.* 2014). In effect, the fruit production of a standard area of Western Amazonian forest cannot be accurately estimated, since the proportion of "female" (fruit-bearing) and "male" (solely pollen-producing) trees remains a mystery in the absence of intensive phenology monitoring efforts.

The vast majority of Western Amazonian tree taxa are also endozoochorous, that is, they produce fruits containing seeds covered in pulp or aril that are designed to attract and be consumed whole by frugivorous vertebrates, which act as seed dispersers by allowing the seed to pass through their digestive tract and emerge intact and pulp-free via defecation or regurgitation. All of the 10 most abundant tree and understory families are dominated by endozoochorous species: Arecaceae, Moraceae, Myristicaceae, Annonaceae, Sapotaceae, Icacinaceae and Lauraceae taxa are exclusively endozoochorous, whereas Malvaceae, Fabaceae and Meliaceae include some anemochorous species but are largely animal-dispersed. By far the predominant animal dispersers are primates, particularly for Moraceae and Sapotaceae. Bats and birds are also very important animal dispersers, and terrestrial rodents are critical dispersers for a small and distinct set of large-seeded species, particularly in the Arecaceae (palms), Fabaceae and Lecythidaceae. *O. parvifolia,* the second-most abundant tree species across the MDDB, is exclusively bat-dispersed, and *I. deltoidea,* the most abundant species, is dispersed primarily by bats and terrestrial rodents. Overall, the predominance of animal-mediated dispersal suggests an important role for plant–animal interactions in the spatial organization of these Western Amazonian floodplain plant communities.

Results from the analyses of intra- and inter-cohort spatial patterns provide several interesting insights into the underlying ecological interactions that structure these plant communities. Significant *intra*-cohort aggregation, i.e. "clumping" amongst conspecific stems (hypothesis 1), was confirmed for 16% of all species examined but more than 80% of species confirmed to a neutral *intra*-cohort spatial distribution pattern. Median percentile ranks of observed NNDs for the three stem cohorts reflected only a modest increase with ontogenetic progression from sapling to adult tree, providing somewhat weak support for the role of resource-based competition and density-dependent thinning of recruit cohorts over time (hypothesis 2). One possible mitigating factor is habitat filtering due to species-specific preferences for microhabitat niches, which can counteract and dilute the effect of negative density-dependent thinning amongst recruit cohorts. It is also likely that the sharpest decline in intra-cohort aggregation occurs during earlier phases of regeneration encompassing the seed-to-seedling transition, such that the spatial distribution pattern of recruit stems $\geq$ 1 m tall more or less represents the long-term intra-cohort spatial organization of the forest tree community, with only modest additional reduction in spatial aggregation over time. Nevertheless, 16% represents a significant proportion of species that have a "clumped" intra-cohort spatial distribution pattern. In comparison, only 10% of all species, pooled across all four sites, had a significantly clumped inter-cohort spatial distribution, and 7%

of all species had a significantly dissociated spatial distribution pattern, with 83% of species confirming to a neutral pattern of inter-cohort spatial organization. In itself, this can be interpreted as ambivalent support of hypothesis 3. However, these results are greatly strengthened when considering the inter-cohort spatial organization of an earlier regeneration phase – between seeds and their parent trees. Post-dispersal seed shadows are highly clumped for the vast majority of canopy tree species, with more than 80–90% of seeds remaining undispersed and aggregated around fruiting adult trees. Previous research conducted at the CC site (Swamy and Terborgh 2010; Swamy *et al.* 2011) has conclusively demonstrated that undispersed seeds located in the vicinity of conspecific adult trees fail to produce saplings and do not contribute to later recruitment stages because they are destroyed by host-specific natural enemies, as described by the J-C hypothesis. Therefore, the predominantly neutral spatial distribution pattern of saplings in relation to locations of conspecific adult trees reflects a striking "de-clumping" of spatial organization over the seed-to-sapling transition. Taking this into account, the predominance of a neutral inter-cohort spatial distribution pattern in relation to $\geq$1 m tall stem recruits and conspecific adult trees provides very strong evidence of dissociation between generations (hypothesis 3), and the overall combined effect of intra- and inter-cohort spatial organization supports a pattern of "displaced self-replacement" (hypothesis 4).

The ecological importance of this inter-cohort spacing mechanism rests in the critical role it plays in maintaining diversity by preventing competitive exclusion and providing habitat space for the recruitment of species that may be inferior competitors for resources. In the absence of this spacing mechanism, an even smaller number of species would dominate available habitat space, crowding out other species and thereby reducing diversity. Given the vast majority of canopy and sub-canopy species in lowland Western Amazonia and other tropical forests that are animal-dispersed, the maintenance of plant diversity in these forests is highly contingent on an intact animal community, particularly populations of large frugivorous vertebrates such as spider monkeys, tapirs, guans, trumpeters and toucans. The ubiquity of both mutualistic and antagonistic interactions between large vertebrates and plants suggests that the dramatic decline of large vertebrate populations observed in tropical forests worldwide in recent decades are likely to have major consequences for compositional and spatial patterns of tropical forest plant communities (Corlett 2007; Dirzo *et al.* 2014; Fa *et al.* 2002; Peres and Palacios 2007; Redford 1992).

Overall, results from this study confirm the long-term, time-integrated outcome of deterministic biotic processes described by the classic J-C hypothesis. Plant-animal interactions (seed dispersal and herbivory) have a strong, lasting influence on diversity patterns at small spatial scales ($\approx$<1 km^2), and on spatial organization within and across generations in lowland Western Amazonian forests. From a conservation standpoint, these results offer strong support for efforts to protect large vertebrate populations in the remaining non-hunted tropical forest sites and to restore them in degraded sites (Brodie and Aslan 2011; Galetti and Dirzo 2013). Without such efforts, the tropical forests of the future could be radically altered with unknown consequences for their stability and the functions and services they provide.

References

Andresen, E. (1999) Seed dispersal by monkeys and the fate of dispersed seeds in a Peruvian rain forest. *Biotropica*, **31**, 145–158.

Brodie, J.F. and Aslan, C.E. (2011) Halting regime shifts in floristically intact tropical forests deprived of their frugivores. *Restoration Ecology*, **20**, 153–157.

Chesson, P.L. and Warner, R.R. (1981) Environmental variability promotes coexistence in lottery competitive systems. *American Naturalist*, **117**, 923–943.

Condit, R., Hubbell, S.P. and Foster, R.B. (1994) Density dependence in two understory tree species in a neotropical forest. *Ecology*, **75**, 671–680.

Condit, R., Ashton, P.S., Baker, P. *et al.* (2000) Spatial patterns in the distribution of tropical tree species. *Science*, **288**, 1414–1418.

Connell, J.H. (1971) On the role of natural enemies in preventing competitive exclusion in some marine animals and in rain forest trees, in *Dynamics of Populations* (eds P.J.D. Boer and G. Gradwell), Centre for Agricultural Publishing and Documentation, Wageningen, The Netherlands, pp. 298–312.

Connell, J.H. (1978) Diversity in tropical rain forests and coral reefs. *Science*, **199**, 1302–1310.

Corlett, R.T. (2007) The impact of hunting on the mammalian fauna of tropical Asian forests. *Biotropica*, **39**, 292–303.

Denslow, J. (1987) Tropical rainforest gaps and tree species diversity. *Annual Review of Ecology and Systematics*, **18**, 431–451.

Diaz-Martin, Z., Swamy, V., Terborgh, J. *et al.* (2014) Identifying keystone plant resources in an Amazonian forest using a long-term fruit-fall record. *Journal of Tropical Ecology*, **30**, 291–301.

Dirzo, R., Young, H.S., Galetti, M. *et al.* (2014) Defaunation in the anthropocene. *Science*, **345**, 401–406.

Fa, J.E., Peres, C.A. and Meeuwig, J. (2002) Bushmeat exploitation in tropical forests: An intercontinental comparison. *Conservation Biology*, **16**, 232–237.

Galetti, M. and Dirzo, R. (2013) Ecological and evolutionary consequences of living in a defaunated world. *Biological Conservation*, **163**, 1–6.

Gentry, A.G. and Terborgh, J. (1990) Composition and dynamics of the Cocha Cashu "mature" floodplain forest, in *Four Neotropical Rainforests* (ed. A.H. Gentry), Yale University Press, New Haven, CT, pp. 542–564.

Grubb, P. (1977) The maintenance of species-richness in plant communities: the importance of the regeneration niche. *Biological Reviews*, **52**, 107–145.

Harms, K.E., Wright, S.J., Calderon, O. *et al.* (2000) Pervasive density-dependent recruitment enhances seedling diversity in a tropical forest. *Nature*, **404**, 493–495.

Hartshorn, G.S. (1978) Tree falls and tropical forest dynamics, in *Tropical Forests as Living Systems* (eds P.B. Tomlinson and M.H. Zimmerman), Cambridge University Press, Cambridge, pp. 613–678.

Hubbell, S.P. (1979) Tree dispersion, abundance, and diversity in a tropical dry forest. *Science*, **203**, 1299–1309.

Hubbell, S.P. (2004) Two decades of research on the BCI forest dynamics plot, in *Tropical Forest Diversity and Dynamism* (eds E.C. Losos and E.G. Leigh Jr.), The University of Chicago Press, Chicago, IL, pp. 8–30.

Hubbell, S.P. and Foster, R.B. (1986) Biology, chance, and history and the structure of tropical rain forest tree communities, in *Community Ecology* (eds M. Diamond and T. Case), Harper & Row, New York, pp. 314–329.

Hubbell, S.P., Condit, R., Foster, R.B. *et al.* (1990) Presence and absence of density dependence in a neotropical tree community (and discussion). *Philosophical Transactions: Biological Sciences*, **330**, 269–281.

Hubbell, S.P., Foster, R.B., O'Brien, S.T. *et al.* (1999) Light-gap disturbances, recruitment limitation, and tree diversity in a neotropical forest. *Science*, **283**, 554–557.

Hurtt, G.C. and Pacala, S.W. (1995) The consequences of recruitment limitation: reconciling chance, history and competitive differences between plants. *Journal of Theoretical Biology*, **176**, 1–12.

Janzen, D.H. (1970) Herbivores and the number of tree species in tropical forests. *American Naturalist*, **104**, 501–528.

Kirkby, C.A., Doan, T.M., Lloyd, H. *et al.* (2000) *Tourism development and the status of Neotropical lowland wildlife in Tambopata, southeastern Peru: recommendations for tourism and conservation*, Tambopata Reserve Society (TReeS), London.

Mangan, S.A., Schnitzer, S.A., Herre, E.A. *et al.* (2010) Negative plant-soil feedback predicts tree-species relative abundance in a tropical forest. *Nature*, **466**, 752–755.

Muller-Landau, H.C., Wright, S.J., Calderon, O. *et al.* (2002) Assessing recruitment limitation: concepts, methods and case-studies from a tropical forest, in *Seed Dispersal and Frugivory: Ecology, evolution and conservation* (eds D.J. Levey, W.R. Silva and M. Galetti), CAB International, Wallingford, UK, pp. 36–53.

Naughton-Treves, L., Mena, J.L., Treves, A. *et al.* (2003) Wildlife survival beyond park boundaries: the impact of slash-and-burn agriculture and hunting on mammals in Tambopata, Peru. *Conservation Biology*, **17**, 1106–1117.

Nunez-Iturri, G. and Howe, H.F. (2007) Bushmeat and the fate of trees with seeds dispersed by large primates in a lowland rain forest in Western Amazonia. *Biotropica*, **39**, 348–354.

Ohl-Schacherer, J., Shepard, G.H., Kaplan, H. *et al.* (2007) The sustainability of subsistence hunting by Matsigenka native communities in Manu National Park, Peru. *Conservation Biology*, **21**, 1174–1185.

Pacala, S.W. and Tilman, D. (1994) Limiting similarity in mechanistic and spatial models of plant competition in heterogeneous environments. *The American Naturalist*, **143**, 222–257.

Peres, C.A. and Palacios, E. (2007) Basin-wide effects of game harvest on vertebrate population densities in Amazonian forests: implications for animal-mediated seed dispersal. *Biotropica*, **39**, 304–315.

Peters, H.A. (2003) Neighbour-regulated mortality: the influence of positive and negative density dependence on tree populations in species-rich tropical forests. *Ecology Letters*, **6**, 757–765.

Pitman, N.C.A. (2010) *An overview of the Los Amigos Watershed, Madre de Dios, Aoutheastern Peru*, Amazon Conservation Association, Washington, DC.

Pitman, N.C.A., Terborgh, J., Silman, M.R. and Nuñez, P.V. (1999) Tree species distributions in an upper Amazonian forest. *Ecology*, **80**, 2651–2661.

Plotkin, J., Potts, M., Leslie, N. *et al.* (2000) Species-area curves, spatial aggregation, and habitat specialization in tropical forests. *J. Theor. Biol.*, **207**, 81–99.

Redford, K.H. (1992) The empty forest. *BioScience*, **42**, 412–422.

Richards, P.W. (1952) *The Tropical Rain Forest: An ecological study*, Cambridge University Press, New York.

Seidler, T.G. and Plotkin, J.B. (2006) Seed dispersal and spatial pattern in tropical trees. *PLoS Biol*, **4**, e344.

Shepard, G.H., Rummenhoeller, K., Ohl-Schacherer, J. and Yu, D.W. (2010) Trouble in Paradise: indigenous Populations, anthropological policies, and biodiversity conservation in Manu National Park, Peru. *Journal of Sustainable Forestry*, **29**, 252–301.

Swamy, V. and Terborgh, J.W. (2010) Distance-responsive natural enemies strongly influence seedling establishment patterns of multiple species in an Amazonian rain forest. *Journal of Ecology*, **98**, 1096–1107.

Swamy, V., Terborgh, J., Dexter, K.G. *et al.* (2011) Are all seeds equal? Spatially explicit comparisons of seed fall and sapling recruitment in a tropical forest. *Ecology Letters*, **14**, 195–201.

ter Steege, H., Pitman, N.C.A., Sabatier, D. *et al.* (2013) Hyperdominance in the Amazonian tree flora. *Science*, **342**, 1243092. DOI: 10.1126/science.1243092.

Terborgh, J. (1983) *Five new world primates: a study in comparative ecology*, Princeton University Press, Princeton, NJ.

Terborgh, J. (1999) *Requiem for Nature*, Island Press, Washington, DC.

Terborgh, J., Pitman, N.C.A., Silman, M. *et al.* (2002) Maintenance of tree diversity in tropical forests, in *Seed Dispersal and Frugivory: Ecology, evolution and conservation* (eds D.J. Levey, W.R. Silva and M. Galetti), CAB International, Wallingford, UK, pp. 1–17.

Terborgh, J., Nuñez-Iturri, G., Pitman, C.A. *et al.* (2008) Tree recruitment in an empty forest. *Ecology*, **89**, 1757–1768.

Tilman, D. (1994) Competition and biodiversity in spatially structured habitats. *Ecology*, **75**, 2–16.

Tobler, M.W. (2008) *The ecology of the Lowland Tapir in Madre De Dios, Peru: using new technologies to study large rainforest mammals*, Texas A&M University.

Tobler, M.W., Carrillo-Percastegui, S.E. and Powell, G. (2009) Habitat use, activity patterns and use of mineral licks by five species of ungulate in south-eastern Peru. *Journal of Tropical Ecology*, **25**, 261–270.

Wallace, A.R. (1878) *Tropical Nature, and Other Essays*, Macmillan & Co, London.

Welden, C.W., Hewett, S.W., Hubbell, S.P. and Foster, R.B. (1991) Sapling survival, growth, and recruitment: relationship to canopy height in a neotropical forest. *Ecology*, **72**, 35–50.

Whittaker, R.H. (1965) Dominance and diversity in land plant communities: Numerical relations of species express the importance of competition in community function and evolution. *Science*, **147**, 250–260.

Wills, C., Condit, R., Foster, R.B. and Hubbell, S.P. (1997) Strong density – and diversity-related effects help to maintain tree species diversity in a neotropical forest. *Proceedings of the National Academy of Sciences of the United States of America*, **94**, 1252–1257.

Wright, S.J. (2002) Plant diversity in tropical forests: a review of mechanisms of species coexistence. *Oecologia*, **130**, 1–14.

9

Bird Assemblages in the *Terra Firme* Forest at Yasuní National Park

Andrés Iglesias-Balarezo, Gabriela Toscano-Montero and Tjitte de Vries

Abstract

The birds of tropical rainforests present the most complex array of ecological interactions. In the Amazon, mixed-species flocks (MSF) have captured the attention of various investigators. Various studies explain that the tendency of birds to form polyspecific groups reduces predatory pressure and increases foraging efficiency. This chapter considers three assumptions suggested by English to classify the flocks: individuals moving at the center of the flock, moving with the flock for more than 30 minutes, and individuals that move with the flock but do not meet the first two assumptions. Canopy flocks presented a less aggressive behavior than understory flocks, as frugivorous birds do not aggressively defend their territories like insectivorous birds do. Habitat type presented a positive correlation with the number of participants in the flocks, suggesting that there is a higher number of species and individuals participating in flocks at zones of *terra firme* forests, contrary to the situation in swamps.

Keywords *canopy flocks; ecological interactions; foraging efficiency; frugivorous birds; insectivorous birds; mixed-species flocks; predatory pressure; terra firme forests; tropical rainforests*

9.1 Introduction

The birds of tropical rainforests present the most complex array of ecological interactions (Pearson 1977a). In the Amazon, mixed-species flocks (MSF) have captured the attention of various investigators (Bates 1864; English 1998; Munn 1985; Tobar *et al.* 2003). Due to the high degree of organization and complexity within the flocks (Graves and Gotelli 1993; Munn 1985; Munn and Terborgh 1979; Powell 1985; Terborgh *et al.* 1990) and because their members interact continuously all through the day, these flocks assemble themselves in very specific ways. Various studies explain that the tendency of birds to form polyspecific groups reduces predatory pressure (Thiollay 1994) and increases foraging efficiency (Greenberg 2000; Munn 1985; Munn and Terborgh 1979).

Forest structure, function and dynamics in Western Amazonia, First Edition.
Edited by Randall W. Myster.

Although there are many studies on MSFs of humid rainforests, most of them focus on understory birds. There are only three studies that describe in detail aspects of natural history, composition and structure of canopy MSFs (de Vries *et al.* 2012; Munn 1985; Piedrahita *et al.* 2003). The information available on these groups is either fragmented (McClure 1967) or only mentioned briefly when speaking about understory MSFs (Baquero 2003; Buitrón-Jurado *et al.* 2003; English 1998; Hutto 1994; Maldonado-Cohelo and Marini 2000; Munn and Terborgh 1979; Terborgh *et al.* 1990).

The work of Munn (1985) describes important findings about flock territoriality and the relationships between core and follower species. He describes the core structure of canopy MSFs in Peru as consisting of a couple or family groups (parents and siblings) that belong to the following six species: *Terenura humeralis, Tolmomyias assimilis, Lanio versicolor, Tachyphonus rufiventer, Tachyphonus luctuosus* and *Hylophylus hypoxanthus*, besides individuals of 53 additional species that join the flocks at different frequencies and at higher or lower densities than core species.

9.2 Methods

This study was made on a 50-ha plot (0 40.506 S; 76 23.984 W) near the Yasuní Research Station (Yasuní National Park) as part of a project financed by REPSOL–YPF and the Pontificia Universidad Católica del Ecuador (PUCE). Most of the fieldwork consisted of observations from along trails and from a bird-watching tower between 6h00–11h30 and 14h30–18h00.

We considered three assumptions suggested by English (1998) to classify the flocks:

1. individuals moving at the center of the flock (where the leader of the flocks and core species forage), at a distance not exceeding 50 m;
2. individuals moving with the flock for more than 30 minutes; and
3. individuals that move with the flock but do not meet the first two assumptions on three different occasions.

Registered species were classified in hierarchies following the definitions suggested by Morse (1970), Munn and Terborgh (1979), English (1998) and Tobar (2006):

- *Leader*: contributes to the formation and disintegration of the flock, and is responsible for guiding other members over long distances to find food and alert them to danger.
- *Core*: Species that spend more than 50% of the day foraging in the flock and generally are members of a unique flock for most of their lives.
- *Facultative*: Species that spend between three and five hours (a day) foraging with the flock.
- *Follower*: Species that spend a few minutes to five hours (a day) foraging with the flock, but are always found at the tail of the flock and never at the center.
- *Occasional*: Species that forage alone at the interior of the forest and whose presence in the flock is ephemeral.

For every flock identified we determined the minimal territory and routes followed throughout the day. The participants were classified according to their feeding guilds, following Terborgh *et al.* (1990): frugivorous, omnivores, insectivores and granivores.

9.3 Results and discussion

In a total of 306.6 observation hours, we found 6 canopy MSFs, although only 4 of them had most of their territories inside the 50-ha plot. The flocks were classified from 1 to 4 according to the order in which they were found inside the plot: flocks 1 and 2 were located near the forest edge where the Maxus road passes, and flocks 3 and 4 were located at the forest interior (Figure 9.1).

We registered between 61 and 86 species of 20 families forming the flocks (Figure 9.2). The number of participants in each flock varied from between 4 and 28 species ($x = 10.5 \pm 1.08$); and from between 7 and 62 individuals ($x = 21.59 \pm 2.48$):

- Flock 1 had 4–28 species ($x = 13.56 \pm 2.40$) and 9–62 individuals ($x = 28.91 \pm 5.56$);
- Flock 2 had 4–14 species ($x = 6.46 \pm 1.59$) and 7–28 individuals ($x = 12.73 \pm 3.44$);
- Flock 3 had 4–26 species ($x = 10.11 \pm 1.50$) and 7–58 individuals ($x = 20.16 \pm 3.17$); and
- Flock 4 had 5–21 species ($x = 11.09 \pm 1.67$) and 10–46 individuals ($x = 24 \pm 8.77$).

The mean territory size was 13.88 ha (±2.05), with flock 1 having the smallest territory while the flock 4 the largest.

The members of the flocks limited their territories by the presence of extensive swamp areas. The formation sites were near water bodies and emergent trees. There is an overlap of 1.8 ha between the territories of flocks 1 and 2 and of 3.6 ha between flocks 3 and 4. We did not find a correlation between the number of species forming each flock and the territory size ($r = 0.797^{NS}$; $p = 0.203$).

The flocks followed several fixed foraging routes during the day. All routes followed *terra firme* forests, avoiding areas where the canopy was interrupted. When encountering a gap in the forest or discontinuous canopy, the flocks never crossed over directly; instead, they circled along the edges until they reached the other end. The flocks never used their whole territory in one day.

According to the occurrence index proposed by Jullien and Thiollay (1998), we registered 12 species considered as obligate participants, representing 12.8% of the 94 total species forming canopy flocks: *Lanio fulvus, Myiopagis caniceps, Myrmotherula ignota ssp. obscura, Euphonia xanthogaster, Saltator maximus, Hylophilus hypoxanthus, T. luctuosus, Tangara velia, Hemithraupis flavicollis, Myiopagis gaimardii, Capito auratus* and *Chlorophanes spiza*. From these, two species assumed the role of leaders according to Munn (1985): *M. caniceps* in flocks 1 and 2, and *L. fulvus* in flocks 3 and 4.

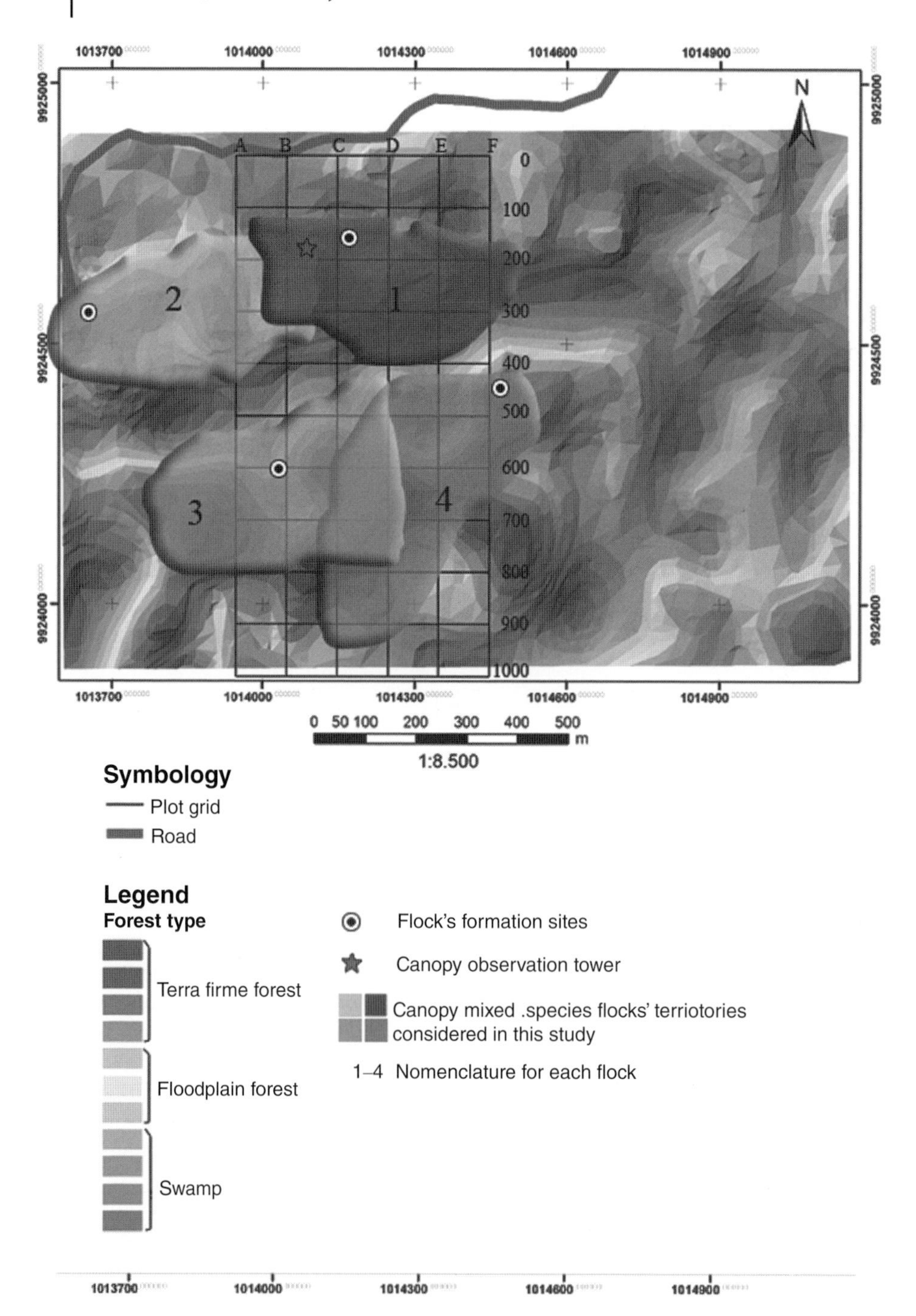

Figure 9.1 Study area with the location of the canopy MSFs studied.

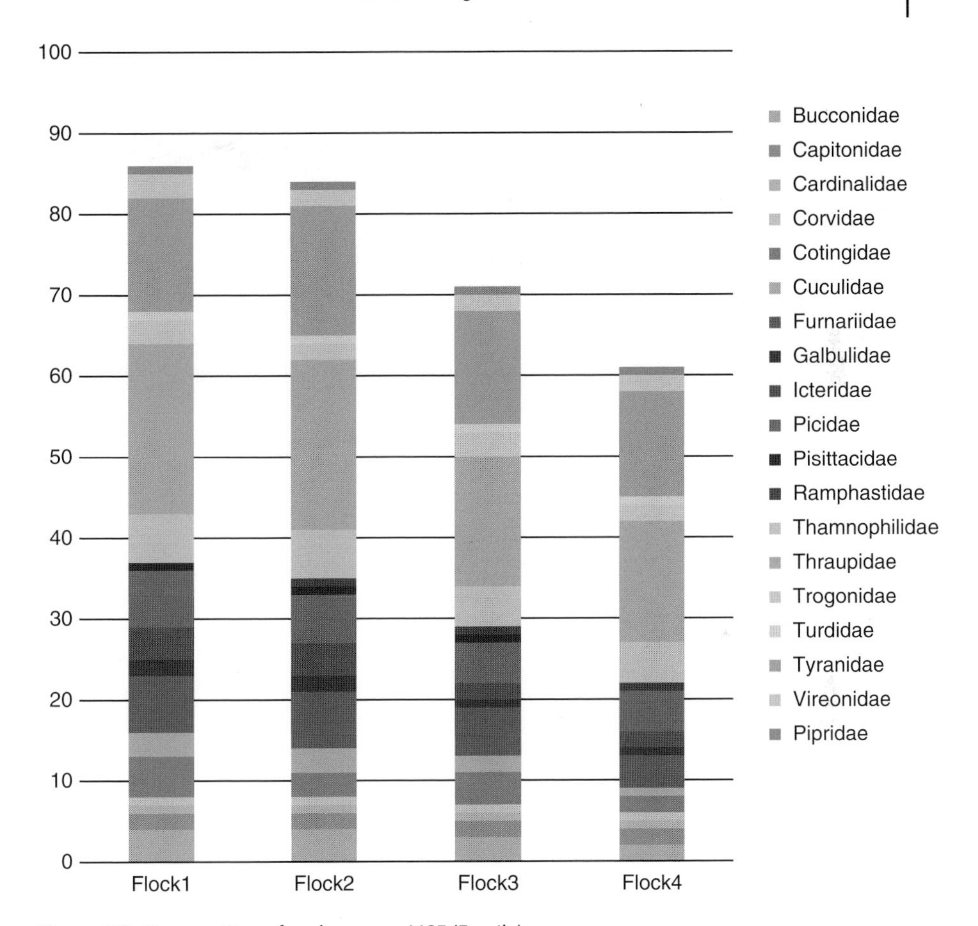

Figure 9.2 Composition of each canopy MSF (Family).

We counted 31 facultative species (33%), 30 followers (31.9%), 13 occasional (13.8%) and 8 accidental (8.5%). The most frequent feeding guild was the insectivorous, with 53 species (53% of the total species), followed by the omnivores.

There were differences in territory size between the canopy MSFs described in this study and of those within the work of Munn (1985). There was also a difference between the understory MSFs of Ecuador and those of other neotropic regions ($\chi^2 = 33.33$, df = 5, $p < 0.001$; $F = 26.163$, df = 57, $p < 0.001$). We classified two groups of flocks according to territory size. The first one includes all of the understory MSFs of Yasuní, Ecuador and Cocha Cashu, Manu, Peru (Terborgh *et al.* 1990), and the second group includes all canopy MSFs (Munn 1985). The territories of canopy flocks were significantly larger than the understory ones (Figure 9.3). The most similar flocks were 1 and 2 with 91% of species shared ($J = 0.91$). These two flocks were located near the road. On the other hand, the most dissimilar flocks were 1 and 4 ($J = 0.58$).

Canopy MSFs have a marked preference for *terra firme* forest ($p < 0.001$**). The flocks remained in this type of forest for more than 70% of foraging

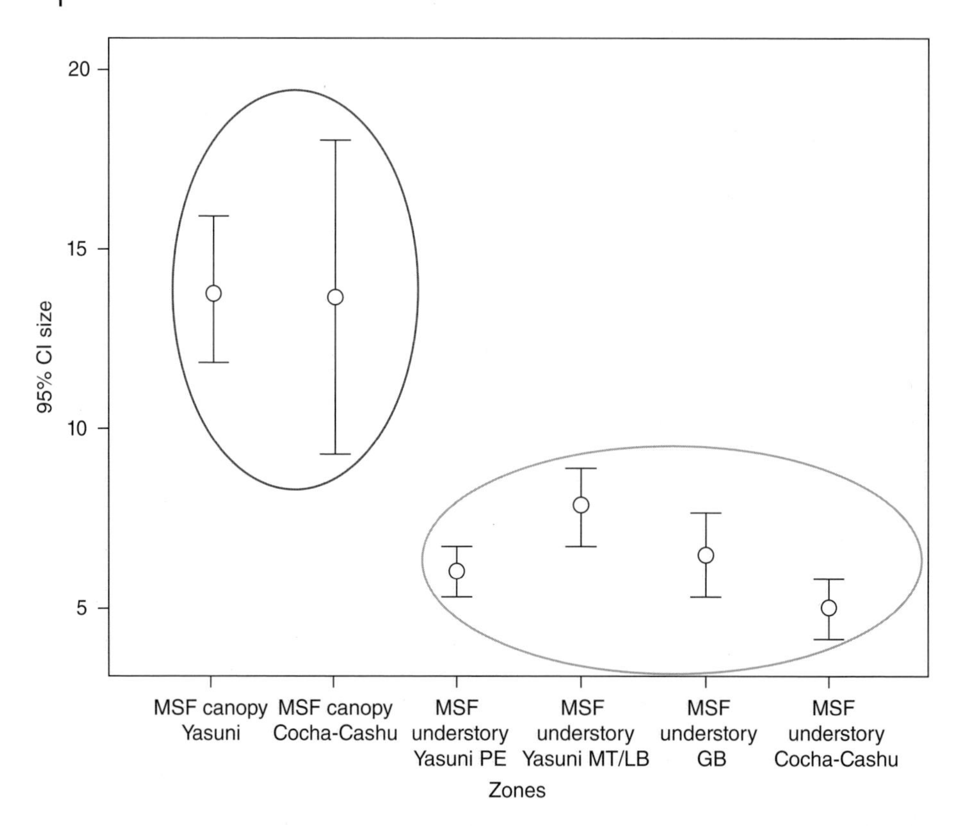

Figure 9.3 Territory size of canopy and understory mixed species flocks. Comparison between six studies. Blue circle: canopy flocks; Green: understory flocks. MSF canopy Yasuní: data presented in this study. MSF canopy Cocha-Cashu: data of Munn (1985) in Manu, Peru. MSF understory Yasuní PE: data from English (1998). MSF understory Yasuní MT/LB: data from Baquero (2003) and Tobar (2006). MSF understory GB: data from Buitrón–Jurado (2005) in Yasuní. MSF understory Cocha Cashu: data from Terborgh *et al.* (1990) in Manu, Peru.

time. Additionally, the flocks preferred to move in zones with closed canopy ($p < 0.001$**), with an altitude superior than 12 m and without gaps produced by tree falls. All the flocks, except flock 2, exhibited this pattern ($p \leq 0.001$**).

The variables that contributed the most to the altitude of foraging were the following: in positive relation canopy continuity ($p < 0.001$**) and vegetation cover ($p = 0.036$*); and in negative association: light intensity ($p = 0.041$*), mean temperature ($p < 0.001$) and foraging intensity ($p < 0.001$).

Daily activity of all flocks began at dawn between 06:07 and 06:35, when the leader couple (*M. caniceps* in flocks 1 and 2 and *L. fulvus* in flocks 3 and 4) vocalized from the tallest trees for 12–18 minutes. The members of some core species responded to the vocalizations and joined the couple to form a group of six to eight individuals of four species. Generally, the first species to join the flock were *H. hypoxantus M. ignota* and *E. xanthogaster.* In flocks 3 and 4, *E. xanthogaster* was replaced by *M. caniceps.*

Once the flock is formed, they move within their territory foraging generally between the treetops. During the morning, other core species, facultative, occasional and accidental species join the groups. Each flock is formed by a couple or family of each species in the case of core and facultative species; for occasional and accidental species the number of participants varied. We identified 68 individuals of 28 species participating simultaneously in the same flock. The peak of activity was between 07:30 and 09:30. Moments before the morning activity ends (at 11:20–12:00), the flock returns to the formation site, moving slowly and quietly. During hot days the daily activity ends earlier (10:30–11:00).

The flocks end their route in the same place they gathered (93% of the time, or 27 of 29 observations), and they stay in the same place until 15:00, at which point they resume their foraging activities. At 17:00, the light in the forest interior diminishes drastically, and the flock ends their activities around this time. Members of the flock spread, beginning with the follower species, then the facultative and lastly the core species. Afternoon foraging was less intensive than in the morning. Flocks disintegrate at approximately 17:30, with each core species occupying a perch near the formation site until the next morning.

We registered five encounters between two flocks where no defensive behavior was observed. Instead, the flocks avoided each other, going in opposite directions while vocalizing. The only aggressive encounter registered was when members of *E. xanthogaster* and *Xenops minutus* joined a flock that already had members of these species. In these cases, the couple of the species forming the flock vocalized and exhibited aggressive behaviors.

Canopy MSFs associated frequently with their understory counterparts. When flocks 3 and 4 associated with understory flocks, the leader was *Thamnomanes caesius,* leader of understory flocks, whereas *L. fulvus* (the leader of canopy flocks) assumed the role of sentinel, giving alarm calls. When the two types of flocks foraged together and passed swamps, vine tangles or temporarily flooded forests, the core of the canopy flocks foraged at the edges or waited until the understory flock ended their foraging activities at these sites. When the understory flocks remained in these areas for more than 30–45 minutes, the canopy flocks separated from the understory flock. The analysis of time spent with understory mixed species flocks showed differences between the flocks near the road and the flocks from the forest interior ($t = -3{,}967{**}$; $p = 0,003$). The flocks of the forest interior foraged longer with understory flocks ($\bar{x} = 85.3$ min.; $N = 10$) than did the flocks whose territories were near the road ($\bar{x} = 5.67$ min.; $N = 3$).

In conclusion, the territories studied for the four canopy MSFs had similar sizes as those studied by Munn (1985) for tropical rainforests. These territories are significantly larger than the understory mixed species flocks, which is why we find fewer canopy flocks in these areas (Munn 1985; Powell 1985). Four MSFs were registered in the same area where eight understory MSFs were described by Buitrón–Jurado (2005).

Canopy flocks presented a less aggressive behavior than understory flocks. This behavior may be due to differences in diet between the flock types, as frugivorous birds do not aggressively defend their territories like insectivorous birds do. Piedrahita *et al.* (2003) suggests that at the canopy there is more substrate where the groups can forage as opposed to in the understory, and thus it is not necessary to defend an abundant resource (Colinvaux, 1993; English 1998; Piedrahita *et al.* 2003).

The closeness of formation sites to water bodies could be explained by the sparser vegetation, particularly around swamps (Clark and Clark 2001), as well as by the greater light intensity inside the forest (Endler 1993). These characteristics make the leader birds more conspicuous and easily located by the other members of the flock. Additionally, sparser vegetation allows the vocalizations emitted by the leader to travel longer distances (Bertelli and Tubaro 2002).

The territory size did not influence the number of participants in each flock, as was the case described by English (1998). Comparing these results with Munn (1985), Yasuní canopy MSFs reported more species participating in the group than those reported in Manu, Peru. The families Thraupidae and Tyrannidae were the most diverse in this study and those made by Munn (1985) and Piedrahita *et al.* (2003). These two-bird families, although phylogenetically distant from each other (Bledsoe 1988; Burns 1998a,b; Carcraft 1981; Raikow and Bledsoe 2000), share social characteristics that allow them to efficiently group forage at intra- and inter-specific levels. The same occurs with the family Thamnophilidae, which is why they can form MSFs more frequently than other bird groups (Buitrón-Jurado *et al.* 2003; Buskirk 1976; Martin and Martin 2001; Munn 1985; Munn and Terborgh 1979; Ridgely and Greenfield 2001a).

Canopy MSFs were composed of insectivorous and omnivorous birds, which was the same situation in the works of Munn (1985), Thiollay (1994) and Piedrahita *et al.* (2003). It is known that arthropods are the most abundant resource exploited by this type of association (Baquero 2003; Buitrón–Jurado 2005; Buskirk 1976; English 1998; Greenberg and Gradwohl 1985; Munn 1985; Tobar 2006).

In this study we recognized 12 core species, 6 species more than the study of Munn (1985). Flocks 1 and 2 located near the road were more similar with each other than with the other flocks, and flocks 3 and 4, which are located inside the forest, were most similar with each other, too. We established that biotic and abiotic variables presented an interdependence. This confirms that the environment influences bird behavior at an individual level as well as at the level of social organization (Bélisle *et al.* 2001; Borges and Stouffer 1999; Graves and Gotelli 1993; Munn 1985; Rosenberg 1990; Walther 2002a; Zimmerman 1966). Temperature greatly influenced the behavior of canopy MSFs, significantly affecting the speed, mean relative foraging height and foraging intensity. At high temperatures, the moving speed and mean relative foraging height decreases, whereas foraging intensity (vocal contact frequency) increases.

It has been proven that at high temperatures, the quantities and the flight speed of insects, the principal food source for these birds, also increases (Taylor 1963). This could be the reason why birds tend to exert a higher foraging effort on hot days. Temperature increases also make the birds descend from the canopy to lower strata of the forest (Iglesias *et al.* 2005; Pearson, 1971). This can be explained by the fact that at higher temperatures, canopy birds tend to descend to the understory in order to preserve body moisture (Pearson 1971; Taylor 1963). The canopy is the vertical strata that receives the most quantities of solar radiation and consequently presents higher temperature peaks, contrary to the sub-canopy and understory where temperatures are more stable (Iglesias *et al.* 2005). For birds in the canopy, it is energetically more taxing to regulate their internal temperatures after being exposed to direct solar radiation, so they descend to lower strata to regulate their body temperature. We determined a negative correlation between temperature and vegetation cover, so canopy MSFs descended to cooler areas with high foliar density, in concordance with Walther (2002a) in Venezuela. In contrast to other variables, light intensity and temperature can change rapidly during the day (Walther 2002a), thus these are factors that may greatly affect bird behavior.

As light intensity rises, canopy flocks descend, especially because at high light intensity levels the risk of being observed by predators also rises (Walther 2002a). Foraging activity of canopy flocks is considerably lower when light intensity is high. During hours of higher temperatures, most of the bird biomass will concentrate at the lower strata of the forest (Pearson 1971). Dry and sunny days reduce the availability of insects and fruits, and the contrast of light to shadow is higher, so insects are harder to catch (Pearson 1977b).

Vegetation cover influenced significantly the relative height of foraging and movement speed. As species in the MSFs entered regions of the forest with a higher percentage of vegetation cover, they foraged at higher altitudes, moved faster and vocalized less frequently. In the canopy, vegetation cover is more important than in lower strata, because most of the biomass concentrates at the understory. Thus, variation of vegetation cover affects behavior in foraging strategies of canopy birds. (Pearson 1975; Walther 2002a). We observed a positive correlation between vegetation cover and the community of canopy birds. In zones with continuous canopy, vegetation cover increases, hence foliar density is higher (Pearson 1975). Birds of MSFs tend to move faster when crossing zones with a higher canopy. In the canopy, the foliar biomass is the highest in a reduced space of about 5–10 m (Clark and Clark 2001; Pearson 1971), and height has an obvious effect in how animals distribute themselves in the forest (Walther 2002b). This height allows each species to establish and take advantage of their own vertical niche (Pearson 1971; Walther 2002b). Nevertheless, given the nature of data used in this study, it was not possible to prove how the height of the canopy determines the vertical distribution of birds. Despite this, it should be noted that we observed a significant negative correlation between canopy height and temperature. Canopy continuity is highly correlated with movement speed, mean relative foraging height and foraging intensity in a negative correlation.

We observed that when MSFs entered zones with discontinuous canopy they foraged at lower altitudes and speed, and presented lower movement rates and vocal contact between its members. This behavior could be explained because in forest clearings, members of the flock are more exposed to predators (Gaddis 1980; Munn 1985). We detected a negative dependence between canopy continuity and temperature. As temperatures rise, the flocks can cross zones with canopy discontinuity. This confirms the fact that when descending to lower strata, the presence of a forest clearing does not constitute a limiting factor to flock movement or foraging (Walther 2002a,b).

Habitat type presented a positive correlation with the number of participants in the flocks, thus suggesting that there is a higher number of species and individuals participating in flocks at zones of *terra firme* forests, contrary to the situation in swamps. Each species has different ecological requirements for height, resources and spatial distribution inside the forest (Graves and Gotelli 1993; Pearson 1971, 1977a,b; Walther 2002a) so many MSFs participants separate from the group when entering habitats that do not offer the resources they require (especially in occasional and accidental participants). This occurs especially in swamps and temporarily flooded forests, since the *terra firme* forest is the habitat that covers the requirements for most species (Graves and Gotelli 1993).

Mean relative foraging height and movement speed are favored in zones with continuous and higher canopies and higher vegetation cover, and they diminish when these variables decrease. Canopy MSFs showed a marked preference for *terra firme* forests over other habitat types (e.g. swamps and temporarily flooded forests), since they spend more time foraging in this type of forest. It appears that *terra firme* forests have more microhabitats that allow the members of the flocks to implement different foraging techniques, reducing competition (Karr 1976; Munn and Terborgh 1979; Robinson and Holmes 1982; Rosenberg (1993).

A bird's predisposition to avoid zones with canopy discontinuity is a well-marked pattern. In induced experiments, birds showed a tendency to take longer routes inside the forest than to cross forest clearings (Bélisle *et al.* 2001; Desrochers and Hannon 1997). Birds use this mechanism to avoid flying between emergent trees and clearings, because it has a higher energetic cost and higher risk of predation (Munn 1985).

There is no doubt that the distance to forest margins, natural or anthropogenic, dramatically alters the composition of animal communities (Hayes and Sewlal 2004; Sekercioglu *et al.* 2002). We registered that the flock leader, *L. fulvus*, was only present in flocks 3 and 4, which are located at the forest interior, whereas *M. caniceps* assumed the role of a leader in flocks 1 and 2, located near the forest edge. Additionally, we observed that flocks from the forest interior shared longer periods foraging with understory flocks than the forest edge flocks did. We can conclude that *L. fulvus* is a species sensitive to habitat modifications, and flocks led by this species are more compatible with understory MSFs than the flocks led by *M. caniceps*. These results are in concordance with English (1998), who explains that encounters between MSFs of the canopy and the understory are well-established behaviors that occur frequently. Analysis of territory limits of the flocks led by *M. caniceps* and *L. fulvus* allowed us to establish a limit for which the edge affects the flock leader by at least 400 m from the forest edge.

Daily activities of the studied canopy mixed species flocks are very similar with other studies (Munn 1985; Piedrahita *et al.* 2003), as well as with understory MSFs (Buitrón–Jurado 2005; English 1998; Jullien and Thiollay 1998; Munn and Terborgh 1979; Tobar 2006): the flock groups, shortly after dawn, responding to calls from the leader species, forage together all day until 17:30 when they disaggregate. Flock territories are as big as the territory of the leader, as was also described in other studies (Munn 1985; Piedrahita *et al.* 2003). In this work, we describe several behaviors not described before:

1. The flocks return to the formation sites to rest at noon, contrary to the understory flocks described by Buitrón-Jurado (2005) and Tobar (2006) that search for vine tangles to rest.
2. Flocks do not use all of the territory on a daily basis. This behavior has been observed in other groups of birds such as tanagers and other frugivorous birds in Costa Rica, where they present a differential selection of the territory according to the season (Loiselle 1988).
3. Formation sites for the flocks were near bodies of water because individuals are easily seen and vocalizations travel farther (Bertelli and Tubaro 2002).

M. caniceps presented a complex vocal pattern when guiding flocks 1 and 2, whereas *L. fulvus* did not have complex vocalizations. However, *L. fulvus* may have displayed the white patches on its wings in order to attract the other members of the flock or other types of visual communication (Ridgely and Greenfield 2001b). These two characteristics are in concordance with the hypothesis that leader species should utilize complex vocalizations or visible marks to fulfill their roles as guiders of the flock (Munn and Terborgh 1979). *Myopagis caniceps* occupied two distinct roles: in flocks 1 and 2 it had the role of leader, meaning it formed the flock, emitted alerts and directed vocalizations. In flocks 3 and 4, it was part of the core species, and it vocalized less frequently since it occupied a role of sub-leader, as described by Munn (1985), English (1998) and Tobar (2006). *L. fulvus* also resigned his leader function while foraging with understory flocks, limiting its role to emitting alarm calls, and acting as a sentinel (Buitrón-Jurado 2005; English 1998). These behaviors confirm the complexity of canopy MSFs in contrast to understory flocks described by Buitrón-Jurado (2005) and Munn (1985).

References

Baquero, L.E. (2003) *Estructura y función de algunas vocalizaciones de Thamnomanes caesius (Thamnophilidae), líder de bandadas mixtas de sotobosque del bosque tropical, en el Parque Nacional Yasuní. Disertación de Licenciatura en Ciencias Biológicas*, Pontificia Universidad Católica del Ecuador, Quito, Ecuador.

Bates, H. (1864) *The Naturalist on the River Amazonas*, vol. **1**, University of California Press, Berkeley/Los Angeles, CA, p. 962(reprinted from the second edition: John Murray: London).

Bélisle, M., Desrochers, A. and Fortin, M.J. (2001) Influence of forest cover on the movements of forest birds: a homing experiment. *Ecology*, **82**, 1893–1894.

Bertelli, S. and Tubaro, P.L. (2002) Body mass and habitat correlates of song structure in a primitive group of birds. *Biological Journal of the Linnean Society*, **77**, 423–430.

Bledsoe, A.H. (1988) Nuclear DNA evolution and phylogeny of the New World nine- primaried oscines. *The Auk*, **105**, 504–515.

Borges, S.H. and Stouffer, P.C. (1999) Bird communities in two types of anthropogenic successional vegetation in central Amazonia. *The Condor*, **101**, 529–536.

Buitrón-Jurado, G., Piedrahita, P., de Vries, T. *et al.* (2003) Dietas y comportamiento de forrajeo de Aves de sotobosque formadoras de bandadas mixtas en el Parque Nacional Yasuní, in *Memorias de las XXVII Jornadas Ecuatorianas de Biología* (eds C.E. Cerón, C.E. and C. Reyes), Publicación de la Sociedad Ecuatoriana de Biología Núcleo de Pichincha, Quito-Ecuador, pp. 9–22.

Buitrón-Jurado, G. (2005) *Competencia interespecífica en aves de bandadas mixtas de sotobosque en el Parque Nacional Yasuní, Amazonía Ecuatoriana*, Disertación previa a la obtención del título de Licenciado en Ciencias Biológicas. Pontificia Universidad Católica del Ecuador, Quito, Ecuador.

Burns, K.J. (1998a) A phylogenetic perspective on the evolution of sexual dichromatism in Tanagers (Thraupidae): the role of male versus female plumage. *Evolution*, **52**, 1219–1224.

Burns, K.J. (1998b) Molecular phylogenetics of the genus *Piranga*: implications for biogeography and the evolution of morphology and behaviour. *The Auk*, **115**, 621–634.

Buskirk, W.H. (1976) Social systems in a tropical forest avifauna. *The American Naturalist*, **110**, 293–310.

Carcraft, J. (1981) Toward a Phylogenetic classification of the recent birds of the world (Class Aves). *The Auk*, **98**, 681–714.

Clark, D.A. and Clark, D.B. (2001) Getting to the canopy: tree height growth in a Neotropical rain forest. *Ecology*, **82**, 1460–1472.

Colinvaux, P. (1993) *Ecology 2*, 2a edn, New York, EE.UU, John Wiley & Sons, Inc.

de Vries, T., Buitrón, G., Tobar, M. *et al.* (2012) Composición, estructura, densidad y aspectos socio-ecológicos de bandadas mixtas de aves de sotobosque y dosel en una parcela de 100 ha, Parque Nacional Yasuní, Amazonia Ecuatoriana. *Revista Ecuatoriana de Medicina y Ciencias Biológicas*, **XXXIII**, 88–123.

Desrochers, A. and Hannon, S.J. (1997) Gap crossing decisions by forest songbirds during the post-fledging period. *Conservation Biology*, **11** (1), 1204–1210.

Endler, J.A. (1993) The color of light in forests and its implications. *Ecological Monographs*, **63**, 1–27.

English, P. (1998) *Ecology of Mixed Species Flocks in Amazonian Ecuador*, PhD thesis. Disertación previa a la obtención del título de Doctor en Filosofía. University of Texas, Austin, TX.

Gaddis, P. (1980) Mixed flocks, accipiters, and antipredator behavior. *The Condor*, **82**, 348–349.

Graves, G. and Gotelli, N. (1993) Assembly of avian mixed-species flocks in Amazonia. *Proceedings of the National Academy of Sciences USA*, **90**, 1388–1391.

Greenberg, R. (2000) *Birds of Many Feathers: The formation and structure of mixed-species flocks of forest birds: On the move: How and why animals travel in groups*, University of Chicago Press, Chicago, IL, pp. 521–558.

Greenberg, R. and Gradwohl, J. (1985) A comparative study of the social organization of Antwrens on Barro Colorado Island, Panama, in *Neotropical Ornithology. Ornithological Monographs, no. 36* (eds P.A. Buckley, M.S. Foster, E.S. Morton *et al.*), American Ornithologist's Union, USA.

Hayes, F.E. and Sewlal, J.A.N. (2004) The Amazon river as a dispersal barrier to passerine birds: effects of river with, habitat and taxonomy. *Journal of Biogeography*, **31** (1), 1809–1818.

Hutto, R. (1994) The composition and social organization of mixed species flocks in a tropical deciduous forest in Western Mexico. *The Condor*, **96**, 105–118.
Iglesias, A., Piedrahita, P., Sánchez, P. *et al.* (2005) Influencia de factores ambientales en la composición, tamaño territorial y altura de forrajeo de las bandadas mixtas de dosel en el Parque Nacional Yasuní. *Resúmenes de las Jornadas Ecuatorianas de Biología*, **XXIX**, p. 23.
Jullien, M. and Thiollay, J.M. (1998) Multi species territoriality and dynamic of Neotropical forest understory bird flocks. *The Journal of Animal Ecology*, **67**, 227–252.
Karr, J.R. (1976) Within- and between-habitat avian diversity in African and Neotropical lowland habitats. *Ecological Monographs*, **46**, 457–481.
Loiselle, B.A. (1988) Bird abundance and seasonality in a Costa Rican lowland forest canopy. *The Condor*, **90**, 761–772.
Maldonado-Cohelo, M. and Marini, M. (2000) Effects of forest fragment size and successional stage on mixed-species flocks in southeastern Brazil. *The Condor*, **102**, 585–594.
Martin, P.R. and Martin, T.E. (2001) Ecological and fitness consequences of species coexistence: a removal experiment with Wood Warblers. *Ecology*, **82**, 189–206.
McClure, H.E. (1967) The composition of mixed species flocks in lowland and sub-montane forests of Malaya. *The Wilson Bulletin*, **29**, 131–154.
Morse, D. (1970) Ecological aspects of some mixed-species foraging flocks of birds. *Ecological Monographs*, **40**, 119–168.
Munn, C.A. (1985) Permanent canopy and understory flocks in Amazonia: species composition and population density, in *Neotropical Ornithology. Ornithological Monographs, no. 36* (eds P.A. Buckley, M.S. Foster, E.S. Morton *et al.*), American Ornithologist's Union, USA, pp. 683–712.
Munn, C.A. and Terborgh, T.W. (1979) Multi-species territoriality in Neotropical foraging flocks. *The Condor*, **8**, 338–347.
Pearson, D.L. (1971) Vertical stratification of birds in a tropical dry forest. *The Condor*, **73**, 46–55.
Pearson, D.L. (1975) The relation of foliage complexity to ecological diversity of three Amazonian bird communities. *The Condor*, **77**, 453–466.
Pearson, D.L. (1977a) A pantropical comparison of bird community structure on six lowland forest sites. *The Condor*, **79**, 232–244.
Pearson, D.L. (1977b) Ecological relationships of small antbirds in Amazonian bird communities. *The Auk*, **94**, 283–292.
Piedrahita, P., de Vries, T., Tobar, M. and Sánchez, P. (2003) Aves de dosel y composición de bandadas mixtas en el bosque tropical del Parque Nacional Yasuní, Ecuador. *Revista de la Pontificia Universidad Católica del Ecuador, Quito, Ecuador*, **71**, 185–199.
Powell, G.V.N. (1985) Sociobiology and adaptive significance of interspecific foraging flocks in the Neotropics. *Ornithological Monographs*, **36**, 713–732.
Raikow, R.J. and Bledsoe, A.H. (2000) Phylogeny and evolution of the Passerine birds. *Bioscience*, **50**, 487–499.
Ridgely, R. and Greenfield, P.J. (2001a) *The Birds of Ecuador: Status, distribution and taxonomy*, vol. **I**, Cornell University Press, Ithaca, NY.

Ridgely, R. and Greenfield, P.J. (2001b) *The Birds of Ecuador: Field guide*, vol. **II**, Cornell University Press, Ithaca, NY.

Robinson, S.K. and Holmes, R.T. (1982) Foraging behaviour of forest birds: the relationships among search tactics, diet, and habitat structure. *Ecology*, **63**, 1918–1931.

Rosenberg, G.H. (1990) Habitat selection and foraging behaviour by birds of Amazonian river islands in northeastern Perú. *The Condor*, **92**, 427–443.

Rosenberg, K.V. (1993) Diet selection in Amazonian antwrens: consequences of substrate specialization. *The Auk*, **110**, 361–375.

Sekercioglu, C.H., Ehrlich, P.R., Daily, G.C. *et al.* (2002) Disappearance of insectivorous birds from tropical forest fragments. *Proceedings of the National Academy of Sciences*, **99** (1), 263–267.

Taylor, L.R. (1963) Analysis of the effect of temperature on insects in flight. *The Journal of Animal Ecology*, **32**, 99–117.

Terborgh, J., Robinson, S., Parker, T.A. III, *et al.* (1990) Structure and organization of an Amazonian forest bird community, in *Ecological Monographs 60(2), 213–238. Neotropical Ornithology. Ornithological Monographs, no. 36* (eds P.A. Buckley, M.S. Foster, E.S. Morton *et al.*), American Ornithologist's Union, USA.

Thiollay, J.M. (1994) Structure, density and rarity in an Amazonian rainforest bird community. *Journal of Tropical Ecology*, **10**, 449–281.

Tobar, M. (2006) *Movimiento, composición y territorio de bandadas mixtas de sotobosque en el Parque Nacional Yasuní. Disertación previa a la obtención del título de Licenciado en Ciencias Biológicas*, Pontificia Universidad Católica del Ecuador, Quito, Ecuador.

Tobar, M., de Vries, T., Piedrahita, P. *et al.* (2003) Composición y territorio de bandadas mixtas de sotobosque en el bosque tropical del Parque Nacional Yasuní, Ecuador. *Revista de la Pontificia Universidad Católica del Ecuador, Quito, Ecuador*, **71**, 167–184.

Walther, B.A. (2002a) Vertical stratification and use of vegetation and light habitats by Neotropical forest birds. *Journal für Ornithologie*, **143**, 64–81.

Walther, B.A. (2002b) Grounded ground birds and surfing canopy birds: variation of foraging stratum breadth observed in Neotropical forest birds and tested with simulation models using boundary constraints. *The Auk*, **119**, 658–675.

Zimmerman, J.L. (1966) Effects of extended tropical photoperiod and temperature on the Dickcissel. *The Condor*, **68**, 377–387.

10

Conclusions, Synthesis and Future Directions

Randall W. Myster

Abstract

This conclusion presents some closing thoughts on the key concepts discussed in the previous chapters of this book. The book focuses on the higher-order complexity and interactively-linked nature of the various ecosystem components within the Western Amazonian rainforest and how it affects us. It discusses the primary importance of permanent plots in the investigation of Neotropic forest structure, function and dynamics. The Amazon is more than just a river, a basin or a rainforest. The book suggests that soil characteristics seem to largely determine floristic composition and flooding largely determines physical structure. It shows a hierarchy of uniqueness among the forest types, with some variation depending on the parameters used, suggested a priority for conservation and management. The book concludes that flooding has significant effects on nutrient availability of Amazonian forest soils by increasing the concentration of some nutrients but decreasing it for others.

Keywords *Amazonian forest soils; flooding regime; floristic composition; Neotropic forest structure; soil characteristics; Western Amazonian rainforest*

10.1 Conclusions

The Amazon is more than just a river, a basin or a rainforest. For our species it provides not just food and various ecosystem services and products, but also nourishes our cultures, our emotions and our psyches. It is literally a part of our shared human consciousness (Figure 10.1). Indeed one of the overarching themes of this book has been the higher-order complexity and interactively-linked nature of the various ecosystem components within the Western Amazonian rainforest and how it affects us.

I argued in Chapter 1 for the primary importance of permanent plots in the investigation of Neotropic forest structure, function and dynamics. And I myself have established 1-ha plots in all five of the most common forest types in Western Amazonia: *terra firme*, white sand and palm (unflooded), *várzea* and *igapó* (flooded). I found in those plots that the most common families were

Forest structure, function and dynamics in Western Amazonia, First Edition.
Edited by Randall W. Myster.

Figure 10.1 La cantina "The Amazons" on the corner of SW 59th Street and south May Avenue in Oklahoma City, Oklahoma (photo courtesy of Nicole Bankert).

Arecaceae, Fabaceae and Clusiaceea, which were also found in all five plots along with Euphorbiaceae, ordinations based on familial data showed that palm forest and white-sand forest were the most different, and that *terra firme* forest, *várzea* forest and *igapó* forest were similar, but ordinations based on physical structure parameters showed a more spread out and individualistic pattern where *várzea* was most different, with palm/*terra firme* and white-sand/*igapó* similar to each other. Furthermore, *terra firme* had the most significant species associations and *igapó* had the least. I found also that *terra firme* and *várzea* had the most negative associations relative to positive associations, and that *igapó*, palm and white-sand had the most positive associations relative to negative associations (author, unpub. data). I concluded that soil characteristics seem to largely determine floristic composition and flooding largely determines physical structure. As forests become more stressed, either by flooding or loss of soil fertility, positive associations became more common, suggesting less competition and more facilitation among the trees. Finally, results show a hierarchy of uniqueness among the forest types, with some variation depending on the parameters used, suggested a priority for conservation and management.

I have also sampled soils and the seed rain in these 5 forest types and found:

1. soil pH of the non-flooded forests was very similar to flooded forests, but became more basic with increased flooding;

2. soil organic matter was lowest in the two non-flooded *terra firme* forests and also increased with flooding;
3. nitrogen (N) was lowest in the palm forest, phosphorus (P) was lowest in *terra firme*-low terrace forest and potassium (K) was lowest in the *terra firme*-high terrace forest;
4. while N decreased sharply with flooding, both P and K increased with length of the flooding period; and
5. for some non-flooded forests there was a correspondence between soil fertility and floristic similarity.

I concluded that flooding has significant effects on nutrient availability of Amazonian forest soils by increasing the concentration of some nutrients but decreasing it for others.

For the seed rain I found:

1. all forests, except black-water tahuampa, contained seeds of tree species that have been sampled in other studies within a forest of the same type, but in all forests there were seeds of several tree species that have not yet been sampled within their forest type;
2. total seed load peaked in the early part of the year – near the end of the rainy season – and then decreased monotonically over the remainder of the year for all forests;
3. species richness was greater in unflooded forests compared to flooded forests, and the largest number of species were found in *terra firme*;
4. seeds were more evenly distributed among species in the unflooded forests compared to the flooded forests; and
5. alpha diversity was much greater in *terra firme* compared to all other forests.

I concluded for the unflooded forests that seed species number and richness increased with soil fertility, but for the flooded forests seed species number and richness decreased with months under water. When taken together, results suggested that for forests across the Amazonian landscape, differences in flooding regime may have a greater effect on both seed rain load and seed species richness than differences in availability of soil nutrients.

Finally field experiments in the five forest types have shown that seed predation took more seeds then either seed pathogenic disease or germination for most seed species, but there were a few species that lost more seeds to pathogens than predators. Germination had the lowest percentage (%) for most species, but again there were a few species where germination % was higher than pathogens, within the unflooded forests. There was the most predation in *terra firme* forest, palm forest lost more seeds to pathogens than predators, and white-sand forest predation levels were between the other two forest types, within the flooded forests, Predation decreased as water went from white to black (at the same inundation levels) and predation decreased monotonically (and pathogens increased monotonically) as months under water increased in black-water forests. There was significantly more seed predation in the unflooded forests compared to the flooded forests, but significantly more germination in the flooded forests compared to the unflooded forests. I concluded that seed predation is the major post-dispersal filter for regeneration in these forests, but pathogenic disease can play a major role, especially in forests that have water in them for long periods each year so that flooding may change those forests dramatically by altering the actions of seed mechanisms and tolerances (author, unpub. data).

Other authors in this book have categorized the Andean–Amazonian interface as embracing the whole area environmentally, but colonial and post-colonial histories have resulted into low levels of cultural, political and economic coherences within the region as a whole. Moreover, considering the spatially variable dynamic environment and the high biodiversity in the region, current scientific comprehension is insufficient to provide a solid basis to support sustainable land allocation and resource use in the region. Another author found 74 species and 21 genera of palm in the eastern Colombian Amazon with 6 different growth forms. The dominant leaf form was the pinnate and only a few species had costapalmate (*Mauritia flexuosa* and *Mauritiella armata*) or palmate leaves (*Lepidocaryum tenue*).

Also authors found that the biogeographic relations of co-existing species are more diverse in the permanently waterlogged wetland habitat than in surrounding forests, suggesting that the main ecological differentiation of taxa with high Andean or low Amazonian biogeographic pattern may be broadly expressed along local environmental gradients in the lowlands. Four canopy mixed-species flocks were composed of 94 bird species. Two of these flocks were headed by *Myopagis caniceps* (the ones closest to a road) and the other two were headed by *Lanio fulvus*. All the flocks preferred *terra firme* forests and avoided temporarily flooded forests and swamps.

10.2 Synthesis

In this volume, and also in my first two books (Myster 2007, 2012a), I, and various other authors, have presented a view of terrestrial ecosystems as plant-centered,

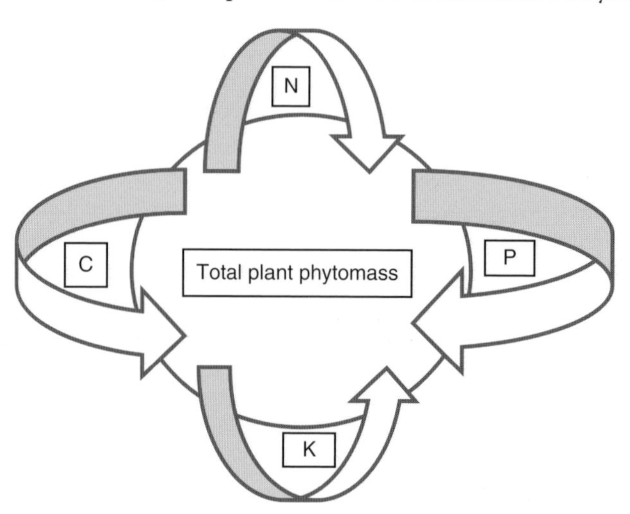

Figure 10.2 A plant-centered view of ecosystems illustrating how important ecosystem components, such as Carbon (C), Nitrogen (N), Phosphorus (P) and Potassium (K), move in and out of the total plant phyto-mass.

where all other components of ecosystems cycle in and out of (or flow through like energy) the total plant phyto-mass (biomass (Myster 2003) and necro-mass together (Figure 10.2). No other component or components of the ecosystem, except the phyto-mass, can assume this central role as a conduit for all parts of the ecosystem (Myster 2001). In addition to largely defining individual cycles, plants also mediate and integrate between cycles. Indeed the phyto-mass conducts the vast majority of an ecosystem's energy and nutrient processing (Grime 1997), where cycles of productivity and decomposition are key.

For example, C in the atmosphere (as CO_2) is taken in by plants and converted to organic molecules (first glucose), which later either decomposes and returns C to the atmosphere (as CO_2) or is consumed by plants or animals, which respire and return C to the atmosphere (as CO_2) and later die and decompose. Plants create C in the soil (as organic matter), which is consumed by the bacteria that take N out of the air and create N compounds that plants take up in their roots (e.g. NO_3-, NH^{4+}). Also plants take up energy from the Sun (as light) and change its form to chemical bonds in organic molecules. When those are consumed by living things, the large molecules are broken down to smaller molecules and energy is released (ultimately as heat), which flows back into outer space.

This framework is in stark contrast to a more common way that ecosystems are conceptualized, where the phyto-mass is presented as just another component, as a "green box". Those modelers ignore species variation, treating all plants as the same (a moss = tree mentality: Myster 2012b). Their focus seems to be on chemistry only, without realizing the central part that vegetation plays in the physics, chemistry and biology of ecosystems. What I am suggesting with this perspective is that all ecologists need to think hard about the structure, function and dynamics of the phyto-mass (at least 90% of which are plants), whenever they study any terrestrial ecosystem.

That phyto-mass is dynamic, changing through time and space, and those changes are determined by the plant–plant replacement process (Myster 2012b) where space is critical due to a plants sessile nature. For Amazon forests, replacements mainly include trees but can also include herbs, shrubs and other plants. I suggest the addition of the plant's phyto-spaces and neighborhood spaces, and replacements that include nine different classes of replacement into investigations of this process (author, unpub. data). This includes replacements that do not result from the death of a plant but from a loss of biomass and/or necro-mass or from release of phyto-spaces and/or neighborhood spaces by other means, where there is plant death but no resulting replacement and involving more than one plant at a time, where the phyto-spaces and neighborhood spaces of those plants may overlap and/or combine in other ways. Fundamentally, I extend the concept of replacement to include space created by plant tissue loss as well as by whole plant mortality.

These spaces are either:

1. the physical space a plants biomass and necro-mass occupies (the phyto-space); or
2. the neighborhood space next to the phyto-space where the plant may influence the nearby environment or be influenced by it (Turkington and Harper 1979).

While the neighborhood space is still being defined, it may change during a plant's life and overlap among different plants. I propose that this new, more comprehensive, model contains nine fundamental replacement classes:

1. that do not result from the death of a plant but from a loss of biomass, or a freeing-up of space by other means, and thus a change in the phyto-space and neighborhood space (replacements none => one and none => many);
2. where there is plant death but no resulting replacement (replacements one => none and many => none); for example, during the thinning phase of forest stand development (Yoda *et al.* 1963), and
3. involving more than one plant at a time (replacements one => many and many => one) whose spaces may combine or overlap.

Finally these replacements should focus mainly on seeds and seedlings. While future models should include the possibility of these kinds of replacements, they should not necessarily assume that any species can be replaced by any other species (George and Bazzaz 1999).

10.3 Future directions

The chapters in this book suggest the continued sampling of all ecosystem components in large forest plots with an emphasis on exploration of interactive links among ecosystem components. Experiments need to be designed to find these links and should have a special focus on the early parts of regeneration (i.e. recruitment of seeds and saplings).

In particular I suggest that researchers:

1. set up permanent plots in various Amazon forest types in different locations and sample them for decades;
2. sample soil and other environmental factors in the plots as well;
3. conduct investigations into the eco-physiology of key tree species found in the plots;
4. determine, with controlled field experiments, the plant–plant replacements which ultimately create forest structure, function and dynamics; and
5. perform restoration experiments needed after logging and conversion to agriculture (Myster 2007).

Finally, I am impressed by the Brazilian and German governments who, some years ago, joined in a long-term partnership to investigate forests in the Central Amazon (discussed in Junk *et al.* 2010). I hope, and would suggest, that the National Science Foundation (NSF) investigates whether such a research partnership could be created between itself and the governments of Peru, Bolivia and Ecuador for the investigation of the Western Amazon. If so, myself and many of the authors in this volume could serve as baseline research scientists and assist in those logistics. In addition to increasing our understanding and helping in the management of the Western Amazon, such a partnership could help address the incredible lack of NSF funding for Amazon research.

References

George, L.O. and Bazzaz, F.A. (1999) The fern understory as an ecological filter: growth and survival of canopy-tree seedlings. *Ecology*, **80**, 846–856.

Grime, J.P. (1997) Biodiversity and ecosystem function: the debate deepens. *Science*, **277**, 1260–1261.

Junk, W.J., Piedade, M.T.F., Parolin, P. *et al.* (2010) *Amazonian Floodplain Forests: Ecophysiology, biodiversity and sustainable management. Ecological Studies*, Springer-Verlag, Berlin.

Myster, R.W. (2001) What is ecosystem structure? *Caribbean Journal of Science*, **37** (1–2), 132–134.

Myster, R.W. (2003) Using biomass to model disturbance. *Community Ecology*, **4** (1), 101–105.

Myster, R.W. (2007) *Post-agriculture Succession in the Neotropics*, Springer-Verlag, Berlin.

Myster, R.W. (2012a) *Ecotones between Forest and Grassland*, Springer-Verlag, Berlin.

Myster, R.W. (2012b) Plants replacing plants: the future of community modeling and research. *The Botanical Review*, **78**, 2–9.

Turkington, R. and Harper, J.L. (1979) The growth, distribution and neighbor relationships of *Trifolium repens*: a permanent pasture. I: Ordination, pattern and contact. *Journal of Ecology*, **67**, 201–218.

Yoda, K., Kira, T., Ogawa, H. and Hozumi, K. (1963) Self-thinning in overcrowded pure stands under cultivated and natural conditions. *Journal of Biology, Osaka City University*, **14**, 107–129.

Index

a
Above-ground biomass (AGB) 4, 9, 11, 12, 13, 86, 87, 91, 92, 93
Acid 5
Adaptations 45, 47, 128, 146
Aerenchyma 129
African oil palm 101
Agriculture 4, 5, 57, 163, 164, 200
Aguarico river 30, 36, 43
Alpha diversity 12, 36, 37, 41, 44, 47, 134
Amazonas river 56, 103, 113, 118, 119
Amigos river 163
Andean–Amazonian 53
Animals 160, 176
Annonaceae 33, 37, 167, 168, 169, 175
Anoxic 128
Apayacuo river 8
Apocynaceae 132
Aquifoliaceae 151
Araguaia-Tocantins river 131
Area de Conservacion Regional Communal de Tamshiyacu-Tahuayo 11
Arecaceae 9, 37, 38, 167, 168, 169, 170, 174, 174, 196
Arthropods 188
Asteraceae 134

b
Bamboo 56
Basal area 9, 10, 11, 12, 13, 86, 93
Bats 175
Begoniaceae 151
Beni river 59
Beta diversity 42, 43, 44, 47, 134, 136
Bimodal 92
Biodiversity 3, 4, 6, 7, 8, 12, 19, 71, 72, 94, 136, 198
Biogeochemical cycles 8
Biogeography 125, 135, 1326
Biomass 4, 199, 200
Birds 175, 181, 189, 198
Bombacaceae 120
Branco river 131
Bray-Curtis 31, 40
Brazil nuts 70
Burning 58
Burseraceae 37, 45

c
Calcium 4, 131
Campanulaceae 151
Cannabaceae 168
Canopy 182, 188, 189, 191
Caqueta river 103, 113, 118, 119
Carbon 3, 8, 67, 86, 87, 94, 133, 199
Carbon dioxide 129
Centro de Investigacion de Jenaro Herrera (CIJH) 13
Chaos diversity index 32, 34, 35
Charcoal 57
Chloranthaceae 151
Chrysobalanaceae 37
Clay 131
Climate change 2, 19, 126, 128
Clumped 171
Clusiaceae 9, 170, 196
Coca 75
Coconut 101

Forest structure, function and dynamics in Western Amazonia, First Edition.
Edited by Randall W. Myster.

Coexistence 8
Cohorts 171, 172
Complexity 7, 19
Condensation 3
Conspecific 175
Cretaceous 3, 126
Cyatheaceae 151

d

Day length 3
Deforestation 7, 36, 67–68, 94, 163
Density-dependence 160, 162, 175
Die-back 135
Diet 187
Disturbances 13, 19, 86, 133, 136, 160
Drought 7, 133, 135
Dry season 14

e

Ecophysiological 8, 200
Ecotourism 71
Ecotypes 129
Edaphic 45, 47, 56, 128, 153
Emergent 190
Endemism 6, 33, 34, 130
Eocene 126
Euphorbiaceae 9, 10, 33, 120, 134, 196
Evaporation 3

f

Fabaceae 3, 9, 18, 33, 37, 38, 120, 167, 168, 169, 175, 196
Facultative 187
Fine roots 133, 134
Fishers alpha 11, 18, 32, 34, 36, 113
Fishers log series distribution 35
Flock 182, 183, 188, 190, 191, 198
Flood pulse 86, 89, 93, 130, 136
Floristics 7, 10, 28, 29, 38, 39, 40, 43, 56, 149, 153
Foraging 187
Fragmentation 6
Frugivores 86, 87, 88, 93
Fulbright foundation 19

g

Gap 11, 12, 16, 17, 70, 87, 160, 174, 183, 186
Generalized least squares analysis 41
Genetic drift 44
Gentry (Alwyn) 148, 149, 150
Geology 43
Germination 13, 119, 130, 134, 197
Gold 72, 164
Gradient 13, 41, 42, 43, 44, 48, 134, 136, 145, 146, 147, 148, 149, 151, 152, 153, 154
Green box 199
Guaviareas river 103, 113, 118, 120
Guild 183

h

Herbarium 30, 36, 37, 38, 103, 104
Herbivory 7, 44, 45, 176
Heterogeneity 16, 43, 160
Hierarchies 182
Higher-order 7, 19, 195
Holocene 9, 11, 127
Humidity 3
Hunter gatherers 57
Hunting 69, 87, 164
Hybrids 45
Hydrochory 130
Hydroelectric 94
Hydropower 136
Hypoxic 128, 147

i

Icacinaceae 167, 168, 175
Igapó 1, 5, 9, 10, 11, 12, 13, 18, 19, 30, 88, 130, 131, 132, 133, 134, 195
Inca 57, 59, 61, 64
Insects 17, 189
Irrigation 136
Isoetaceae 151

j

Janzen-Connell hypothesis 160, 161, 176
Jesuit 62
Jurassic 126
Jurua river 88

l

Land use 7

Lauraceae 33, 38, 151, 167, 168, 169, 175
Leaf area ratio (LAR) 16, 18
Leaf mass ratio 16
Lecythidaceae 9, 37, 132, 175
Leguminosae 9
Lenticels 129
Linear regression analysis 11
Linked 195
Litter 131, 132
Logging 4, 68, 69, 70, 74, 163, 164, 200
LTER 12, 17

m

Madre de Dios river 56, 59, 75, 162, 163
Magnesium 131
Magnoliaceae 151
Malvaceae 9, 167, 168, 175
Maranon river 56
Mass effects hypothesis 12
Melastomataceae 33, 38, 43, 151, 169, 170
Meliaceae 120, 167, 168, 169, 170, 175
Mining 72, 163, 164
Miocene 45, 54, 126, 127
Moraceae 33, 37, 38, 120, 167, 168, 169, 170, 174, 175
Morona river 40
Mycorrhizal association 8
Myristicaceae 37, 167, 168, 169, 170, 174, 175
Myrtaceae 9, 38, 132, 151

n

Necro-mass 199
Negro river 131, 135
Napo river 36, 43, 75
Natural gas 66
Neighborhood space 199, 200
Net primary productivity 3
Niche 39, 135, 146, 149, 153, 154
Nitrogen (N) 4, 14, 15, 131, 132, 197, 199
Non-metric multi-dimensional analysis 31
NSF 200

o

Oil 66, 71, 74
Oligarchies 28, 120, 173
Oxygen 129

p

Paleo- 56, 92, 127, 128, 136
Paleocene 126
Paleogene 126
Palm 1, 5, 9, 10, 13, 14, 15, 19, 29, 30, 56, 101, 102, 103, 104, 111, 113, 118, 119, 120, 121, 195, 196, 198
Paradigms 159
Pastaza river 75
Peat 147
pH 3, 14, 44, 196
Phenology 8, 87, 88, 89, 92, 130, 175
Phosphorus (P) 3, 4, 7, 14, 15, 131, 132, 197
Photosynthetic active radiation (PAR) 134
Phyto-mass 199
Phyto-space 199, 200
Piperaceae 169, 170
Plasticity 44
Pleisotcene 3, 30, 128
Plot 8, 9, 13, 16, 17, 28, 30, 32, 33, 89, 161, 162, 164, 173, 182, 183, 195, 200
Pneumatophores 129
Pollen 126
Pollination 47, 160
Polygonaceae 132
Polypodiaceae 151
Potassium (K) 3, 4, 14, 15, 131, 197
Precipitation 11, 102
Primulaceae 151
Principal component analysis 149
Production 132
Pteridophytes 43
Putumayo river 43, 75

q

Quartz 5, 9, 14
Quaternary 127, 128

r

Rainfall 14, 15, 88
Rainy season 3, 11, 88, 197
Rao's quadratic entropy 149
Recruitment 7, 160
Regeneration 159, 160, 174, 175, 200
Relative growth rate 16
Remote sensing 8
Replacement 19, 149, 161, 199, 200
Resprouting 129
Root/shoot ratio 16
Rosaceae 151
Rubber 64, 65
Rubiaceae 9, 10, 33, 38, 120, 151, 169

s

Sabalillo forest reserve 8
Sand 44, 131
Sandstone 39, 40, 56
Sapling 171
Sapotaceae 37, 38, 167, 175
Sedimentation 133, 134
Seed dispersal 7, 41, 44, 86, 87, 92, 94
Seed dormancy 130
Seedling 8, 16, 18, 200
Seed pathogens 13, 18, 197
Seed predation 13, 160, 197
Seed rain 15, 196, 197
Silt 131
Soil bulk density 10, 11
Soil fertility 197
Soil organic matter 14, 129, 132, 197, 199
Soils 4, 7, 200
Solanaceae 151
Speciation 44, 45
Species diversity 4, 56, 145, 159, 161
Species richness 11, 44, 47, 197
Specific leaf area 16
Strata 165, 189
Sustainable 66, 69
Swietenia macrophylla 70

t

Tambopata river 164
Tapajos river 131
Tefe river 131
Temperature 14, 15, 88, 102, 188
Terra firme 1, 4, 5, 6, 8, 9, 10, 11, 12, 13, 14, 15, 16, 17, 18, 19, 29, 30, 33, 36, 41, 86, 88, 89, 91, 92, 102, 104, 111, 113, 118, 119, 127, 129, 130, 132, 136, 145, 146, 150, 151, 183, 185, 190, 195–198
Thamnophilidae 188
Thraupidae 188
Thymeleaceae 151
Tiputini river 16, 17
Transects 103
Trombetas river 131
Turnover 12, 44, 45, 47, 149
Tyrannidae 188

u

Uatuma river 131
Ucayali river 56, 59, 119, 120, 121
Understory 182, 187, 188
Urticaceaea 10, 134

v

Varzea 1, 5, 6, 13, 16, 17, 18, 19, 30, 85, 86, 88, 89, 91, 92, 93, 94, 127, 128, 130, 131, 132, 133, 134, 135, 195, 196
Violaceae 120
Vocalizations 191

w

Water 4, 7
White sand 1, 5, 9, 10, 13, 14, 15, 19, 30, 36, 37, 40, 45, 56, 132, 195–197
Wood densities 86, 87, 134

x

Xingu river 131

y

Yasuní 9, 15, 16

z

Zingiberaceae 43